Challenges in Physics Education

This book series covers the many facets of physics teaching and learning at all educational levels and in all learning environments. The respective volumes address a wide range of topics, including (but not limited to) innovative approaches and pedagogical strategies for physics education; the development of effective methods to integrate multimedia into physics education or teaching/learning; innovative lab experiments; and the use of web-based interactive activities. Both research and experienced practice will feature prominently throughout.

The series is published in cooperation with GIREP, the International Research Group on Physics Teaching, and will include selected papers from internationally renowned experts, as well as monographs. Book proposals from other sources are entirely welcome.

Challenges in Physics Education addresses professionals, teachers, researchers, instructors and curriculum developers alike, with the aim of improving physics teaching and learning, and thereby the overall standing of physics in society.

Book proposals for this series my be submitted to the Publishing Editor: Marina Forlizzi; email: Marina.Forlizzi@springer.com

More information about this series at http://www.springer.com/series/16575

Jenaro Guisasola · Kristina Zuza
Editors

Research and Innovation in Physics Education: Two Sides of the Same Coin

 Springer

Editors
Jenaro Guisasola
Donostia Physics Education Research
Group (DoPER), Department of Physics
Applied, Gipuzkoa School of Engineering
University of the Basque
Country (UPV/EHU)
Donostia-San Sebastian, Spain

Kristina Zuza
Donostia Physics Education Research
Group (DoPER), Department of Physics
Applied, Gipuzkoa School of Engineering
University of the Basque
Country (UPV/EHU)
Donostia-San Sebastian, Spain

ISSN 2662-8422   ISSN 2662-8430   (electronic)
Challenges in Physics Education
ISBN 978-3-030-51184-5   ISBN 978-3-030-51182-1   (eBook)
https://doi.org/10.1007/978-3-030-51182-1

This Springer imprint is published by the registered company Springer Nature Switzerland AG
The registered company address is: Gewerbestrasse 11, 6330 Cham, Switzerland

Introduction

This book presents selected contributions from International Research Group on Physics Teaching (GIREP) and Multimedia in Physics Teaching and Learning (MPTL) Conference, held in Donostia-San Sebastian, Spain, from 9 to 13 July 2018. The GIREP community combines the efforts and interests of two communities, researchers in physics education and physics teachers. This book presents a broad field of research and innovation that ranges from basic research to innovations in classroom teaching. The research and proposals include different educational levels from Primary to University. Together they make a strong contribution to knowledge on physics teaching and learning.

This edited volume features 19 manuscripts the plenary lectures, dialogues, workshops and communications selected from the most outstanding papers presented during the Conference. The theme of the conference was "Research and Innovation in Physics Education: Two Sides of the Same Coin". The papers highlight relevant aspects of the relationships between research and innovation in the teaching of physics. These studies provide new knowledge to improve learning processes and instruction. The book is of interest to teachers and researchers committed to teaching and learning physics based on evidence.

The organization of the Conference would not have been possible without the help of many people. In particular, we would like to thank Prof. Marisa Michelini, President of GIREP, for her constant support. We thank the members of the Local Organizing Committee and the reviewers for their dedication and commitment to this event, without them the proven quality of the contributions would not have been possible. We hope that this book provide the reader with a contemporary vision of the teaching and learning process of physics

Jenaro Guisasola
Kristina Zuza

Contents

Experiments as Building Blocks of Knowledge

Gorazd Planinšič

Abstract When scientists are constructing new knowledge, they design their own experiments to observe new phenomena, to test their hypotheses or to apply acquired knowledge. When the same scientists present their new ideas to expert colleagues, they describe outcomes of their experiments, show how these outcomes support their models and how these ideas can be applied. Which of these two situations is closer to the current view of how teaching–learning process in the school should look like: constructing new knowledge or presenting the knowledge? Once we realize that teaching by telling is terribly inefficient and that learning only occurs when students are actively engaged in the learning process, we also realize that the traditional role of experiments in the schools and in the textbooks is no longer useful. In my talk, I will describe how changes in my own views about the role of physics experiments helped me to see new features of the experiments that would otherwise remain hidden to me. I will show examples of experiments that are used to achieve active engagement of students in the learning processes, experiments that served as a resource for designing new type of problems and experiments that were used as a research tools in PER.

1 Introduction

I am deeply grateful to GIREP community for recognizing the value of my work by awarding me the 2018 GIREP Medal. Incidentally, it is exactly 20 years since I have attended my first GIREP conference in Duisburg, Germany in 1998. However, all these years alone would not help if I did not meet several exceptional people whose ideas, experiences and critiques helped me to come closer to the goal that I am trying to reach—and this is, how to better use experiments in teaching and learning. My deep thanks go first to Eugenia Etkina, who introduced me to ISLE—a theoretical framework that tremendously improved my work and my teaching during the past

G. Planinšič (✉)
University of Ljubljana, Ljubljana, Slovenia
e-mail: gorazd.planinsic@fmf.uni-lj.si

J. Guisasola and K. Zuza (eds.), *Research and Innovation in Physics Education: Two Sides of the Same Coin*, Challenges in Physics Education,
https://doi.org/10.1007/978-3-030-51182-1_1

ten years. I would also like to thank Leoš and Irena Dvorak, Marisa Michelini, Josip Slisko and Laurence Viennot for sharing ideas and giving me opportunity to learn through the projects that we created together. I would also like to thank my Slovenian colleagues Mojca Čepič and Aleš Mohorič for their work and support in achieving recognition of physics education as a research field of physics in Slovenia.

2 Short History

Experimenting and finding out how things work fascinated me from my childhood. I did my Masters degree and the PhD in experimental MRI on nonconventional systems that we had to build from scratch. The time I spent working in the MRI field was a great lesson in how to design experiments and how to test different ideas in physics. In 1996, I joined my friend Miha Kos in his endeavours to establish the first Slovenian hands-on science centre, called The House of Experiments. Looking back I think it was the work with the House of Experiments that made me switch from MRI to physics education. I applied for the open position for a leader of physics education program at the faculty for mathematics and physics at the University of Ljubljana and got the job. The main mission of the program was and still is to prepare high-school physics teachers for Slovenia. Being aware that I was a novice in physics education field, I started attending physics education conferences, visiting PER groups around the world (such as PEG by Lillian McDermott), collaborating with colleagues with similar interest abroad and publishing. For many years, I was passionately learning about experiments, building them and using them in the classroom but also designing my own, study physics behind them, use them in the classroom with students-future physics teachers, with in-service teachers, improve them based on their responses, and wrote papers of what I learned. Some papers came to the front pages of renowned journals, some were translated to other languages, and some have been cited over 25 times so far (Planinšič 2001, 2004, 2007; Planinšič and Kovač 2008). I was invited to give workshops and talks and became active in international physics education community. I co-organized GIREP Seminar in 2005 in Ljubljana, became secretary of GIREP and later a chair of Physics Education Division at European Physical Society (EPS PED). Years spent in EPS PED were very productive. At the meeting of EPS PED in CERN in 2008, Laurence Viennot proposed to collect good examples of activities with the right balance between hands-on and minds-on. Christian Ucke, late Elena Sassi and myself immediately volunteered to join this project, which was later called more understanding with simple experiments (MUSE). Today, the MUSE webpage contains more than 50 freely available activities for physics teachers (https://www.eps.org/members/group_content_view.asp?group=85190&id=187784).

These were my first ten years working in physics education. I expected that all this new knowledge and many great experiments will also help me to become a better teacher, but this did not happen to the extent that I hoped it would. I was good in specific examples that challenged students to apply acquired knowledge in a new situation, but I was not able to transfer these ideas to situations in which students need to construct new concepts and basic ideas for themselves—a crucial task for every teacher. In short, I was missing a framework for teaching and learning that would help me better prepare future physics teachers and that would resonate with my own interest for experimental physics.

3 ISLE

As I implied earlier, I eventually found a suitable theoretical framework, named ISLE (Etkina and Van Heuvelen 2007). But, before describing the meaning of the acronym and the structure of the framework, let me present a simplified example that illustrates how ISLE works.

On a hot summer day, you pour ice-cold water into a glass and ask students to say what they observe using only terms that are familiar to them. They quickly notice the water drops on the outside of the glass on the part where water fills the glass (Fig. 1). The next question is to work in groups to come up with several explanations of where this water came from and write down the explanations. Usually, students or other participants come up with the following explanations: (1) the water from the glass seeped through the glass wall; (2) the water escaped from the top of the glass and landed on the outside; (3) water on the outside of the glass did not come from the water in the glass, it came from the air outside. Once all the explanations are listed and shared the next step is to ask—what do we do next? Usually, one of the students says: We need to test them. How do we test explanations? The students propose to do more experiments. But what experiments to do? Here, the instructor helps them: Let us come up with new experiments whose outcomes we can predict using every explanation and then compare the outcomes with the predictions. Table 1 shows the testing experiments that the students come up with predictions based on each explanations, outcomes and final judgment (Table 1). Note that it does not matter that many students know the "right" answer. The point is to devise multiple explanations and think how to test them. After all ideas except (3) are ruled out by

Fig. 1 Ice-cold water in a glass—observational experiment

Where do these tiny water drops come from?

Table 1 Different explanations for the observed water drops, typical proposals for testing experiments, corresponding predictions (at crossroads), outcomes of the testing experiments and judgements

	Testing exp. 1: Use dry and cooled glass (put glass in a fridge)	Testing exp. 2: Use different cold liquid (ex. oil) *Assumption*: there is no water in oil	Testing exp. 3: Original experiment + weigh glass (initial/final)	Testing exp. 4: Original experiment + cover glass *Assumption*: cover stops water
Explanation 1: Water from the glass seeped through glass wall	No water outside glass	No water outside glass	$m_f = m_i$	Water on outside glass
Explanation 2: Water escaped from the glass and landed on the outside glass	No water outside glass	No water outside glass	$m_f \leq m_i$	No water outside glass
Explanation 3: Water from air collected on the wall outside glass	Water on outside glass	Water on outside glass	$m_f > m_i$	Water on outside glass
Outcomes	Water on outside glass	Water on outside glass	$m_f > m_i$	Water on outside glass
Judgments	Reject 1, 2	Reject 1, 2	Reject 1, 2	Reject 2

testing experiments, students are asked if there is any practical use for this knowledge. They brainstorm and come up with ideas such as drying humid places by extracting water from air, collecting drinking water from air in the dessert and other real-life applications.

This example illustrates the essential idea of the interactive method of teaching, Investigative Science Learning Environment (ISLE) (Etkina and Van Heuvelen 2007; Etkina 2015). The theoretical framework of ISLE was originally developed by Eugenia Etkina from Rutgers University USA and later enriched by collaboration with Alan Van Heuvelen. ISLE helps students learn physics in ways similar to how physicists work, by engaging students in investigative processes that mirror the practice of physics. The example above shows how students can learn something new by being engaged in an investigative process in which they follow a logical progression from observing simple experiments to devising multiple explanations, testing them and applying them for practical purposes. ISLE process can be schematically represented by a flow diagram (Fig. 2). It is important to note that the process is neither linear nor cyclic. ISLE is a framework within which you can organize inquiry activities in a way that resembles the way physicists do research. Another two key features of ISLE are the use of multiple representations that help students develop the reasoning skills and collaborative work that allows all students to participate in the process.

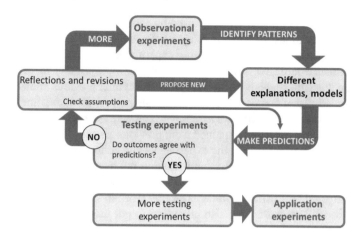

Fig. 2 ISLE process

The roles of experiments in ISLE resemble the roles they have in physics; they can either be observational, testing or application experiments. Eugenia came to this categorization of experiments by studying the roles that experiments played in history of physics. But the ISLE process does not only reflect the way physics knowledge was built historically. It turns out that the ISLE process also closely resembles the way how scientists solve experimental problems in real time, as we showed in qualitative research few years ago (Poklinek Čančula et al. 2015).

4   Changes

What first attracted my attention about ISLE was a new way of looking at the experiments. At that time, I was already aware that seeing experiments as demonstration experiments and laboratory experiments does not help much in achieving active learning where students are active participants in all steps of the learning process, not passive observers, but I did not know a better alternative. The role of demonstration experiments is mostly to demonstrate the theory presented by a teacher, and the role of traditional laboratory experiments is mostly to engage students in processes in which they verify the same theory experimentally. ISLE showed me an alternative way how to look at experiments that I accepted immediately (Etkina et al. 2002). Roles of experiments in ISLE are not rigid (e.g., a testing experiment can turn into an observational experiment), and most importantly, these roles give an experiment an active character "by default". For me, looking at the experiments as observational, testing or application experiments was a first major step towards my goals to improve my teaching and to achieve active learning. The next important step for me was to realize the didactical value of the logical flow that resembles the process used by expert physicists. ISLE helps students learn and build a coherent knowledge by slowing down the process typically used by scientists, unpacking it and making it explicit. For example, articulating the predictions for the outcomes of the testing experiments before performing the testing experiments may not be something that expert scientists would always do (they do this in their mind without articulating), but insisting on this with students allows them to connect their previous knowledge with a new knowledge that comes as a result of the activity. At the same time, students' predictions can serve as a formative assessment and give teachers opportunity to help students close some gaps that remained from previous lessons.

A different way of looking at the experiments and adopting the new logical flow was not the only important change that I made in my teaching. I made other major changes in our program for preparing future high-school physics teachers, all stimulated or inspired by ISLE. (1) Students strictly work in small groups using whiteboards to show their work (this also allows us to put more emphasis on multiple representations). (2) I ask students for predictions only when they can base them on explanations under test (not on intuition). (3) Students are given opportunities to improve. They can revise their work without penalties. (4) Every meeting ends with students' reflections (what did I learn today… as a physicist, as a teacher).

New view on the role of experiments and the process of learning gave also new momentum to my work. I started to collaborate with ISLE creators and developing new ISLE-based materials (Planinšič and Gojkošek 2011; Planinšič and Etkina 2012; Etkina et al. 2013; Planinšič et al. 2014; Etkina and Planinšič 2015). I also revised my previous ideas, looking at them through new "glasses". Table 2 shows the first few steps of an activity that I developed before I learned about ISLE and how I changed it after that. I choose the activity that utilizes computer scanner for learning about relative motion (Planinšič et al. 2014).

If you compare the two activities in Table 2, you will notice two important things. First, in the ISLE version, students are significantly more active than in the original version (compare the amount of italicized texts). Second, students working in groups are able to come up with an unconventional graphical representation (note that the scanner head is modelled as a point while the car is modelled as an extended object). This last came as a complete surprise for me. Initially, I was sure that students will not be able to come up with this representation on their own, without guiding.

5   Modern Devices in Introductory Physics Course

Success of scanner activities made us realize that modern (or contemporary) devices can be productively used in introductory physics courses even when the students do not understand in details how a device works as long as they have an opportunity to test their ideas within the context in which the device is used. Eugenia and I expanded this idea to a framework for using modern devices in introductory physics courses while developing a systematic series of ISLE-based activities for learning about light emitting diodes (LED) (Planinšič and Etkina 2014).

The framework consists of three different ways of using a modern device:

Table 2 Changing an old activity into ISLE activity

Initial activity (first 6 steps)	"ISLE-ised" activity (first 6 steps)
1. Give students a clear explanation how a scanner captures an image (image is captured line-by-line, while the scanner head is moving at constant speed) 2. Tell students that we will place a toy car on the scanner and that we will be using special $x(t)$ graph (see below) to study motion of the car and the scanner head in various situations. Discuss the graph elements 	1. Place a car on the scanner window. Let students observe scanning process (with open scanner) and the result (car is at rest) 
3. Show scanned image of a car for $v_{car} = 0$ 	2. Ask students to come up with different explanations how the scanner captures the image *(Usual explanations are* *E1: Line-by-line; E2: From the whole window using a long time exposure)* 3. Ask students to suggest experiments to test their explanations *(Most frequent idea: remove the car from the scanner when the scanner head is at about the middle of the car.)* 4. Ask students to make predictions for the outcome of the testing experiment based on the explanations under test *(If E1, then half of the car on the image will be missing. If E2, then the whole car will be on the image but less bright)* 5. Let students observe the outcome of the testing experiment and make judgments 
and ask how the image will change if car moves during scanning process so that $v_{car} < v_{scanner head}$ *(Students make a prediction. In most cases, their prediction is correct—image will be longer)* 4. Perform the experiment 	
5. Ask students to compare the outcome with the prediction. If they do not match, ask students to resolve what was wrong in their reasoning *(In most cases, the outcome matches the prediction)* 6. Ask students to think how we can use the $x(t)$ graph introduced earlier to make the prediction about the size of the car on the scanned image *(Students correctly interpret the graph and find out how the graph can be used to predict the relative size of the scanned image)* 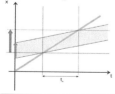	*(Students reject E2 and accept E1 as correct explanation)* 6. Ask students to represent the motion of the scanner head and the car (as seen by the observer in the lab reference frame) on one graph. Ask them to come up with the type of the graph that will also allow predicting the relative length of the scanned image of the car *(Students, working in groups, consistently come up with the $x(t)$ graph as shown below. After students make predictions, they perform experiments and compare outcomes with their predictions)* 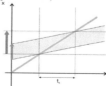

Only first six steps are shown. Typical student responses are written in brackets and italicized. The complete ISLE activity is described in Planinšič et al. (2014)

1. as a black box,
2. to learning how a modern device works,
3. to learn new physics using the knowledge of how the device works.

The proposed framework is not meant to be "theoretical" framework but rather a guide that will help teachers to think of how to incorporate modern devices (MD) in introductory physics curriculum.

5.1   Using a Modern Device as a Black Box

Using a MD as a black box allows students to get familiar with certain properties of a system of interest without going into the physics of the device itself. However, even though we do not seek explanations for how the device works, black boxes offer several opportunities for connecting, comparing and contrasting features of a MD with other devices or phenomena already familiar to the students. Black box activities can be seen as a first step in getting personal experience with a new device or a piece of technology, moving from unknown, abstract to known, concrete. For example, we use LEDs as light sources when teaching optics, as current indicators when teaching electricity and magnetism or as blinking light sources when studying motion in kinematics (Planinšič and Etkina 2015a). Each of these cases offers an opportunity to discover some special features of an LED and compare those with the features of some other devices already familiar to students (such as incandescent light bulbs).

5.2   Learning About How a Modern Device Works

We can engage students in activities in which they learn the basics of how a MD works using knowledge of unit-relevant physics to come up and test different explanations related to the operation of MD. The explanations can be either causal (relating cause and effect) or mechanistic (explaining the mechanism behind the phenomenon). We start with qualitative investigations and then proceed to quantitative description and testing explanations. For example, after learning basics about DC circuits including Ohm's law, students can start qualitative investigation by trying to make a green LED glow using two 1.5 V batteries and do the same for a small incandescent light bulb. We can encourage students to propose causal explanations for the observed behaviour of LEDs and then to propose testing experiments to test these ideas (Etkina and Planinšič 2014). Students then proceed to quantitative investigation by measuring the current-versus-voltage characteristic $I(\Delta V)$ of an LED and a light bulb and realize that their measurements are consistent with what they found out earlier in the qualitative investigation. If we decide to go further, we can engage students in the next activity that allows them to create an image about the structure of an LED, thus creating a need to know and preparing them for "time for telling" (Schwartz and Bransford 1998) abut a p-n junction (Etkina and Planinšič 2014). Observing an LED immersed in glycerine or in a silicon oil using a microscope is a fascinating experience for every student (see Video 27.1 at https://media.pearsoncmg.com/aw/ aw_etkina_cp_2/videos/content/videos.php). Students will learn that the heart of an LED is a cake-like object that consists of a thick layer covered by a thin layer of a different material that glows when an LED is connected to a battery.

5.3   Learning New Physics Using Knowledge of How MD Works

The idea of this step of the framework is to build on the knowledge about a MD that students obtained through the activities in the previous step with one of the two goals: either to learn about a new phenomenon (that is a part of the curriculum) in a new context or to deepen understanding of physics and broaden the knowledge by exploring new phenomena. Activities in this step are usually suitable for advanced high-school courses and introductory courses for physics and engineering majors but also for courses for prospective physics teachers and professional development programs. For example, once students have learned the basics of LEDs as described in the previous step, we can engage them in an activity in which they discover fluorescence (a new phenomenon) while investigating how a white LED works (Planinšič and Etkina 2015b). We assume the students have also learned the basics of colour light mixing, including the concept of complementary colours and the basics of wave optics.

6   New Types of Problems

I had a privilege to join Eugenia and Alan in writing the second edition of an algebra-based introductory physics textbook, accompanying collection of activities called active learning guide (ALG) and instructor's guide (IG) that are entirely based on ISLE framework (Etkina et al. 2019a, b, c). In the last two years, I spent a great deal of time designing and testing new types of problems for the textbook and ALG. By new, I mean the problems that go beyond the problems that require from students to find one correct numerical answer. Or in other words, problems that help student develop critical thinking and other specific competences that are important for working in science. The types of the problems are not my invention (several types of these problems were already in the first edition of the textbook)—what I am trying to show here is again, how new way of looking at experiments and observing students, while solving experimental problems allowed me to create new problems of these types. I will present only two examples, but interested readers are encouraged to look for the rest of the types in the textbook, ALG and IG mentioned above.

6.1 "Tell All" Problems

Here is the problem [variation of the problem 75 on page 115 in our textbook (Etkina et al. 2019a)].

> You push a 1.7-kg book along a table and let go. The book comes to a stop after a short distance. Figure 3 shows the acceleration-versus-time graph of the book as recorded by an accelerometer that was fixed on the book (the mass given above includes the mass of the accelerometer). List as many quantities as you can that can be determined based on the data given in the problem and determine three of them. Indicate any assumptions that you made.

I came up with this problem when I observed students (future physics teachers) learning how to use a remote controlled accelerometer. Later, we realized that the same measurements can be obtained with an accelerometer that is integrated in a mobile phone using one of the free applets. Looking at the graph through "ISLE glasses" gave me an idea that reversing the problem (making students go from the graph back to the description of motion) can turn a usual measurement into a detective story.

"Tell all" problems allow students to go into various depths depending on their level of preparation, giving opportunities to all students to feel successful. Less prepared students will only determine the peak value of the acceleration or the average negative acceleration. More prepared students will determine the coefficient of kinetic friction between the book and the table. The same problem offers also challenging questions for your most advanced students. At which moment (point on the graph) does the hand lose contact with the book? Or can we determine from the graph at which moment the hand made a first contact with the book? Another important feature

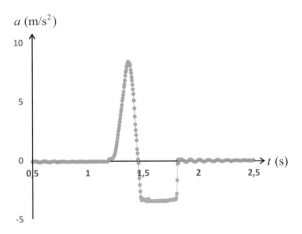

Fig. 3 The acceleration-versus-time graph for the book that is pushed along the table and slows down to a stop

of "tell all" problems is that they require the students to think about assumptions and to think how assumptions affect the result. Assumptions are crucial in any scientific work and yet we hardly ever speak about them in high-school or university physics courses.

6.2   Evaluate Reasoning or Solution

Here is the problem [problem 32 on page 81 in our textbook (Etkina et al. 2019a)].

Students Lucia, Isabel and Austin are investigating how snow stops a dropped 500-g lemon juice bottle. In particular, they are interested in how the force exerted by the snow depends on the age of the snow. They take high-speed videos of the bottle, while it is sinking into the snow, taking their first set of measurements 4 days after fresh snowfall and the second set of measurements 2 days later. After analysing the videos frame by frame, they plot a graph that shows how the velocity of the bottle from the moment the bottle touches the snow changes for both types of snow (see Fig. 4).

They each explain their results as follows:

Lucia: The six-day snow exerts a larger force on the bottle because it stops the bottle in a shorter time.

Isabel: The time taken to stop the bottle does not say much about the force. The six-day snow exerts a larger force on the bottle because the slope of its $v_y(t)$ graph is steeper.

Austin: We cannot compare the forces exerted by the snow because the initial velocities are different. We need to repeat the experiments and make sure we always drop the bottle from the same height.

Explain how each student reached her/his conclusion and decide who (if anyone) is correct. Indicate any assumptions that you made.

Fig. 4 The velocity-versus-time graph for a bottle that slows down to a stop in the snow

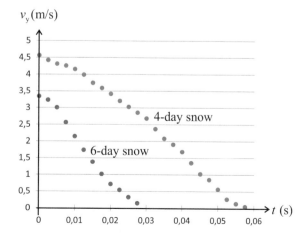

I was working on the chapter on Newton's laws, searching for fresh ideas from everyday life that would involve accelerated motion. Being trapped in the house by the snow, I came up with an entertaining idea, to analyse how a falling object slows down while sinking in a snow. My intuition was telling me that the acceleration during slowing down will be changing at different rate (probably largest at the end) which would limit the usefulness of the problem for the introductory level. I decided not to rely on my intuition and do the measurement. I was very surprised when I found out that the speed was changing almost linearly. At first, I was happy. I found a new context, and I got measurements that allowed me to design a problem for the chapter on Newton's laws. But in two days, the excitement went away. The material that I had still did not allow me to design a problem that goes beyond the problems that require from students to find one correct numerical answer. Then, it occurred to me if I can make another measurement at different conditions, this may allow us to design a richer problem. The final form of the problem (students discussing different ideas) evolved through the discussion with Eugenia.

This type of problems requires from students to make an effort to understand other's ideas and critically evaluate their reasoning. While doing this, students have to recognize productive ideas (even when they are embedded in incorrect answers) and differentiate them from the unproductive ideas. The awareness of assumptions is also important in this type of problems. When the force exerted by the snow depends on the velocity of the bottle, Austin's suggestion becomes very reasonable.

7   Summary

At the end, let me summarize what I think are two most important messages that I would like to send.

- Looking at the experiments as observational, testing and application experiments is the first step towards helping students to think like scientists and teachers to make progress towards active learning. The next and crucial step is embracing the ISLE process.
- Modern or contemporary devices can be integrated in physics curriculum without overloading it. The framework presented in this paper is simple to use and can be applied to any device and any course. As the gap between the research and technological achievements on hand and the mandatory physics content on the other hand is growing, the integration of modern devices into introductory courses will become vital in future.

References

Etkina E (2015) Millikan award lecture: students of physics—listeners, observers, or collaborative participants in physics scientific practices? Am J Phys 83:669–679

Etkina E, Planinšič G (2014) Light-emitting diodes: exploration of underlying physics. Phys Teach 52:212–218

Etkina E, Planinšič G (2015) Defining and developing critical thinking through devising and testing multiple explanations of the same phenomenon. Phys Teach 53:432–437

Etkina E, Van Heuvelen A (2007) Approach to learning physics. In: Redish EF, Cooney PJ (eds) Research-based reform of University Physics, vol 1. Retrieved from: www.compadre.org/per/per_reviews/media/volume1/isle-2007.pdf

Etkina E, Van Heuvelen A, Brookes DT, Mills D (2002) Role of experiments in physics instruction—a process approach. Phys Teach 40:351–355

Etkina E, Planinšič G, Vollmer M (2013) A simple optics experiment to engage students in scientific inquiry. Am J Phys 81:815–822

Etkina E, Planinsic G, Van Heuvelen A (2019a) College physics: explore and apply, 2nd edn. Pearson, New York

Etkina E, Brookes D, Planinsic G, Van Heuvelen A (2019b) Active learning guide for college physics: explore and apply, 2nd edn. Pearson, New York

Etkina E, Brookes D, Planinsic G, Van Heuvelen A (2019c) College physics. Instructor's guide, 2nd edn. Pearson, New York

https://media.pearsoncmg.com/aw/aw_etkina_cp_2/videos/content/videos.php. Retrieved 25 Feb 2019

https://www.eps.org/members/group_content_view.asp?group=85190&id=187784. Retrieved 26 Feb 2019

Planinšič G (2001) Water-drop projector. Phys Teach 39:76–79

Planinšič G (2004) Color mixer for every student. Phys Teach 42:138–142

Planinšič G (2007) Project laboratory for first-year students. Eur J Phys 28:S71–S82

Planinšič G, Etkina E (2012) Bubbles that change the speed of sound. Phys Teach 50:458–460

Planinšič G, Etkina E (2014) Light-emitting diodes: a hidden treasure. Phys Teach 52:94–99

Planinšič G, Etkina E (2015a) Light-emitting diodes: solving complex problems. Phys Teach 53:291–297

Planinšič G, Etkina E (2015b) Light-emitting diodes: learning new physics. Phys Teach 53:210–216

Planinšič G, Gojkošek M (2011) Prism foil from an LCD monitor as a tool for teaching introductory optic. Eur J Phys 32:601–613

Planinšič G, Kovač J (2008) Nano goes to school: a teaching model of the atomic force microscope 2008. Phys Educ 43:37–45

Planinšič G, Gregorčič B, Etkina E (2014) Learning and teaching with a computer scanner. Phys Educ 49:586–595

Poklinek Čančula M, Planinšič G, Etkina E (2015) Analyzing patterns in experts' approaches to solving experimental problems. Am J Phys 83:366–374

Schwartz DL, Bransford JDA (1998) Time for telling. Cogn Instr 16:475–522

Active Learning Methods and Strategies to Improve Student Conceptual Understanding: Some Considerations from Physics Education Research

Claudio Fazio

Abstract Active learning methods and strategies are credited to be an important means for the development of student cognitive skills. This paper describes some forms of active learning common in Physics Education and briefly introduces some of the pedagogical and psychological theories on the basis of active learning. Then, some evidence for active learning effectiveness in developing students' critical cognitive skills and improving their conceptual understanding are examined. An example study regarding the effectiveness of an Inquiry-based learning approach in helping students to build mechanisms of functioning and explicative models, and to identify common aspects in apparently different phenomena, is briefly discussed.

1 Introduction

Active learning (AL) methods and strategies have received considerable attention over the last several years and are commonly presented in the literature as a credible solution to the reported lack of efficacy of more "traditional" educative approaches (Cummings 2013). There is today a wide consensus in admitting that much of the knowledge taught in schools and universities by following traditional educative approaches, that are often focused on a one-way transmission (i.e., from the instructor to the learner) of essential principles, concepts, and facts, is not easily retrievable in real-life contexts. Research has shown that a possible cause of this is the abstract and decontextualized nature of traditional education, which often ignores the interdependence of situation and cognition. When learning and context are separated, knowledge itself is seen by learners as the final product of education rather than a tool to be used dynamically to solve problems (Herrington and Oliver 2000).

AL methods and strategies are credited to improve student conceptual understanding in many fields, including physics (e.g., Georgiou and Sharma 2015; Sharma

C. Fazio (✉)
Physics Education Research Group, University of Palermo, Palermo, Italy
e-mail: claudio.fazio@unipa.it

© The Editor(s) (if applicable) and The Author(s), under exclusive license to Springer Nature Switzerland AG 2020
J. Guisasola and K. Zuza (eds.), *Research and Innovation in Physics Education: Two Sides of the Same Coin*, Challenges in Physics Education,
https://doi.org/10.1007/978-3-030-51182-1_2

et al. 2010; Hake 1998; Redish and Smith 2008). For these reasons, AL has gained strong support from teachers and faculties looking for effective alternatives to traditional teaching methods. However, some remain skeptical about its real efficacy and see it as one more in a long line of educational fads (Prince 2004). Many also express doubts about what AL is and how it can be considered different from traditional education. Particularly, they claim that their teaching methods can already be considered "active," as homework assignments and, in many cases, laboratories are part of them. Adding to the confusion, many teachers and faculties do not always understand how the most common forms of active learning differ from each other. In some cases, they are not inclined to comb through the educational literature for answers.

In this paper, after the first definition of AL, we discuss some of the pedagogical and psychological theories on its basis. Second, we distinguish among different types of AL strategies most frequently discussed in the Physics Education Research (PER) literature. Basic core elements are identified for each of these separate types to differentiate among them, and a brief review of the literature regarding AL effectiveness in developing specific student skills and in improving their conceptual understanding and motivation is performed. Finally, an example study regarding the effectiveness of a specific approach to active learning (i.e., the Inquiry-based one) in helping students to build mechanisms of functioning and explicative models, and to identify common aspects in apparently different phenomena developed at the University of Palermo, Italy, is discussed.

2 Active Learning

The term "active learning" was first introduced by the English scholar Reginald W. Revans in his pioneer studies on Action Learning (Revans 1982). Active learning (AL) can be considered a form of learning in which teaching strives to strongly involve students in their learning, encouraging them to do things and reflecting on things they are doing. According to Bonwell and Eison (1991), *"in active learning, students participate in the process, and students participate when they are doing something besides passively listening."*

2.1 Learning Theories

Active learning methods and strategies are mainly based on the constructivist theory of learning, which describes the way people may effectively acquire knowledge and learn. The theory suggests that humans construct knowledge and meaning from their experiences, and its first formalization is generally attributed to Jean Piaget. He suggested that individuals construct new knowledge from their experiences through the processes of *assimilation* and *accommodation*. When the experiences are aligned with the individuals' internal representations of the world, they *assimilate* (i.e., incorporate) them into an already existing framework without changing that framework. In contrast, when individuals' experiences contradict their mental representations of the external world, they must reframe these representations to fit the new experiences, *accommodating* the new ideas in the pre-existing schemas inside their minds. The individual's knowledge develops from the continuous interaction between the two processes. He/she assimilates the characteristics of the environment in the mental schemes suitable to contain them; when needed, he/she adapts his/her mental schemes to new experiences, thus creating a continuous and balanced circle between assimilation and accommodation.

So, learning is a dynamic process comprising successive stages of adaption to reality during which learners actively construct knowledge by creating, testing, and reframing their theories of the world.

Very soon, cognitive scientists agreed on the fact that learning is deeply influenced by social interaction, as the learner arrives at his/her version of the truth influenced by his/her background, culture or embedded worldview. The sociocultural perspective in learning was fostered, among the others, by Bandura (1977) and Vygotsky (1986). Historical developments and symbol systems, such as language, logic, and mathematical systems, are inherited by the learner as a member of a particular culture, and these are learned throughout the learner's life. This also stresses the importance of the nature of the learner's social interaction with knowledgeable members of society. Without social interaction with other, more knowledgeable people, it is impossible to acquire the social meaning of important symbol systems and learn how to utilize them. People develop their thinking abilities by interacting with other people and the physical world. From the social constructivist viewpoint, it is thus important to take into account the background and culture of the learner throughout the learning process, as this background also helps to shape the knowledge and truth that the learner creates, discovers, and attains in the learning process

In his studies, Vygotsky also clarified that each learner brings experience to the learning situation, i.e., existing knowledge and preconceptions. These include beliefs formed through various experiences that can deeply influence new knowledge construction. This aspect is particularly relevant for the learning of scientific disciplines that aim at describing and explaining situations coming from the real world. In this case, it is highly probable that the learner comes to the learning environment with spontaneous models (i.e., naive ideas about how the world works). They are often strongly validated by his/her real-life experience and are hardly influenced by traditional educative approaches.

The constructivist models of human learning led to the development of a theory of *cognitive apprenticeship*. This theory holds that to foster learning effectively, the teacher should take into account the implicit processes involved in carrying out complex skills. Cognitive apprenticeships are designed, among other things, to bring these tacit processes into the open, where students can observe, enact, and build representations of the world (i.e., models) and practice them with help from the teacher. This approach is supported by Bandura's theory of modeling, according to which the learner must be motivated to learn, must have access to and retain the information presented, and must be able to reproduce the desired skill accurately.

Part of the effectiveness of the cognitive apprenticeship model comes from *learning in context* and is based on theories of *situated cognition.* Cognitive researchers argue that cognitive apprenticeships are less effective when skills and concepts are taught independent of their real-world context and situation. Learning and cognition are fundamentally situated.

Jerome Bruner is often credited with originating the idea of *discovery learning* in the 1960s, but ideas very similar to Bruner's ones can be found in earlier writers (e.g., Dewey 1916). Discovery learning can be considered heavily inspired by the constructivist approach to education and is based on an approach to learning based on inquiry and discovery. Bruner argues that practice in discovering for oneself teaches one to acquire information in a way that makes that information more readily viable in problem-solving. This philosophy later became the discovery learning movement of the 1960s, whose mantra suggests that people best *learn by doing.* Discovery learning takes place in problem-solving situations where the learner draws on his own experience and prior knowledge. It is an instruction method through which students interact with their environment by exploring and manipulating objects, wrestling with questions and controversies, performing experiments, and building descriptive and explicative models of the observed regularities. The final result is a *self-direction* of learning that can have the best results on the learner's conceptual understanding and appropriation of transversal skills that are likely to play an important role during all his/her life.

The literature on constructivist models of human learning, so, suggests that useable knowledge is best gained in active learning environments, which feature the following characteristics (e.g., Herrington and Oliver 2000):

1. provide *authentic contexts* that reflect the way knowledge will be used in real life
2. provide *authentic activities* that may also be complex, ill-defined problems and investigations
3. provide access to *expert performances* including modeling of processes
4. provide *multiple roles and perspectives* that allow the learner to search for alternative solution pathways
5. support *collaborative construction of knowledge* allowing for the social construction of knowledge
6. promote *reflection* to enable abstractions to be formed and promoting metacognition
7. promote *articulation* to enable tacit and/or common-sense knowledge to be made explicit
8. provide *coaching* and *scaffolding* by the teacher at critical times
9. promote the authentic assessment of learning within the tasks, that reflects the way knowledge is assessed in real life.

2.2 Theories on Learners' Mind and Psychological Types

To foster authentic self-directed learning, both the instructor and the learner need to be at least aware of how the mind learns. Many studies in this field, like to ones regarding the concept of the *Growth Mindset* introduced by Dweck (2006), highlighted that students who have this understanding demonstrate higher scholastic and academic success. According to Dweck's idea, Growth Mindset students are generally aware of how their learning may happen. They firmly believe that extra focused effort and motivation may improve their "intelligence" (Atkinson and Feather 1966) and may allow them to acquire expert skills from experience. Conversely, students that have a "fixed mindset" often do not believe that any effort can help them to improve their skills and understanding. In many cases, they simply duplicate the teacher's portrayal of critical thinking and problem-solving, not fostering their conceptual knowledge and appropriation of essential transversal skills.

So, the Growth Mindset prepares the students to assume responsibility for self-directed knowledge acquisition from their experiences. According to Ericsson (2004), the process found to be most effective in acquiring expert skills from experience is known as *Deliberate Practice*. This is a particular type of practice that is purposeful, systematic, and performed at progressively more challenging levels (Mayer 2008). While regular practice might include mindless repetitions, Deliberate Practice requires focused attention and is conducted with the specific goal of improving performance. A fundamental aspect of this process is that the student can develop *self-awareness* of his/her points of strength and weakness. This also allows the teacher to help focus practice that can be repeated at different levels of difficulty to improve a skill that is found as weak by both him/her and the student him/herself. Emphasis on self-awareness in Deliberate Practice is likely to play a role in the effectiveness of the Growth Mindset (Pelley 2014a).

If Deliberate Practice is to be applied to learning, then some knowledge of theories on *learning styles* (e.g., Barbe et al. 1979; Fleming and Mills 1992) and on *psychological types* (Jung 1971) should be possessed by both teachers and students. Notably, a student cannot understand the purpose of a learning strategy if he/she does not understand how learning happens.

According to the VARK model (Fleming and Mills 1992), a refinement of the previous VAK model of learning styles (Barbe et al. 1979), four different channels for learning exist, namely visual, auditory, read/write, and kinesthetic:

Visual learners learn by seeing. They prefer information obtained by visual representations such as graphs, maps, and displays. They frequently move hands while talking and tend to look upward when thinking (Pritchard 2009).

Auditory learners learn by listening. They prefer repetition, summaries, and benefit from discussions, lectures, stories, and podcasts. They tend to tilt their heads and use eye movements when concentrating or recalling information (Pritchard 2009).

Read/write learners prefer information displayed as words. Their preference emphasizes text-based input and output—reading and writing in all its forms but especially manuals, reports, essays, and assignments.

Kinesthetic learners prefer to do something to learn. They privilege interactions within the learning environment and especially with their bodies. They will easily recall events or information attached to experience or the feelings of a physical event. They learn best through field trips, physical activity, manipulating objects, and touch.

According to Carl G. Jung's theory of psychological types (Jung 1971), people can be characterized:

i.   by their preference of general attitude: *Extraverted* or *Introverted*,
ii. their choice of one of the two functions of perception: *Sensing* or *Intuition*,
iii. their preference for one of the two functions of judging: *Thinking* or *Feeling*.

These three Jungian areas of preferences are dichotomies (i.e., bipolar dimensions where each pole represents a different preference). Jung also proposed that in a person, one of the four functions above is dominant—either a function of perception or a function of judging.

The first criterion, Extraversion/Introversion, signifies the source and direction of a person's energy expression. An extravert's source and direction of energy expression are mainly in the external world, while an introvert has a source of energy mainly in their inner world.

The second criterion, Sensing–Intuition, represents the method by which someone perceives information. Sensation-type persons generally trust tangible, concrete information, i.e., information that can be gained/understood by the five senses. Intuition-type persons tend to trust information that is more abstract and that can be associated with other information (either remembered or discovered by seeking a wider context or pattern).

The third criterion, Thinking–Feeling, represents how a person processes information. Thinking-type persons tend to decide things from a detached standpoint, measuring the decision by what seems reasonable, logical, causal, consistent, and matching a given set of rules. Feeling-type persons tend to come to decisions by associating or empathizing with the situation, looking at it "from the inside" and weighing the situation to achieve, on balance, the greatest harmony.

Psychological types seem at first to have little to do with both learning and skill development. However, a review of the ideas first described by Jung and of the subsequent literature shows that not only is skill involved but that many skills are involved (Myers et al. 1998). For example, when the Sensing and Intuitive preferences are compared, they can be at first seen as mutually exclusive opposites. In this case, however, a powerful opportunity for skill development is missed. When viewed instead as intellectual skills, the mental processes that underlie the preference become mutually beneficial. Thus, according to Pelley (2014b), a preference for Sensing and Intuition can be referred to as a "learning style," and Sensing and Intuitive functions can be referred to as learning "skills." A preference for Sensing or Intuition, so, is better understood as a *comfort zone* where the student spends most of his/her thinking time because he/she is more at ease with it. However, the student can apply Deliberate

Practice to develop the opposite function as a thinking skill. This is possible to some extent because any person can adapt the use of both Sensing and Intuition to each learning situation regardless of his/her preference. Everyone can use both "skills" while having only one "preference" (Pelley and Dalley 2008).

A teacher can help the student to become aware of when and how to use his/her opposite perception function and guide his/her development into a self-directed mentality. Thus, according to Pelley (2014a), a learning style is a preference, not a limitation. An example can be found in the process of throwing a stone. Throwing with the person's dominant hand is natural and done virtually unconsciously because motor function becomes automated over time (i.e., the person is in his/her comfort zone). Switching hands to the less preferred one requires that the person switches the attention to the *mechanics* of throwing with the less used limb. The process is at first "child-like" and underdeveloped (Pelley 2014a); however, with practice, the non-dominant hand use can be improved. If an instructor aids Deliberate Practice, this can happen at a more rapid pace.

In the same way, a student can develop the less used components of their cognition. Thus, an Intuitive-type person would have the Sensing function as his/her off-hand and vice versa for the Intuitive type. Further, if an Intuitive-type person wants to develop his/her Sensing skills by Deliberate Practice, he/she would undertake learning activities that require active incorporation of detailed facts into the integration that he/she does intuitively. An example is the addition of detailed facts to the tips of the branches in a concept map. Indeed, as described by Pelley (2014a), concept maps are excellent Deliberate Practice modalities. Sensing-type students can improve their ability to discover integration, and Intuitive-type ones can construct branches where they can hang all of those elusive details (Pelley 2014a).

2.3 Different Types of Active Learning

As we have already said, any instructional method that engages students in the learning process can be defined as "active." While this could include traditional activities such as homework or laboratory, in practice, AL refers to activities that are introduced into the classroom. Different AL methods have been discussed in the literature. For some of them, we here highlight some core elements that can help us to differentiate among them.

Active lecturing is a form of lecturing during which the lecturer pauses periodically to allow students to clarify their notes with classmates and to say what they think explicitly. The lecturer can also actively engage students using direct questions regarding the lecture. The core element of active lecturing is the possibility for students to discuss and reflect on the subjects of the lecture during its development.

Collaborative learning refers to pedagogical situations in which students work together in small groups toward a common goal (Smith and MacGregor 1992). The main idea here is to replace individualistic learning with learning based on student group work. So, the core element of collaborative learning is the emphasis on student interactions (also in a competitive way), rather than on learning as a solitary activity.

Cooperative learning can be defined as a structured form of group work where students pursue common goals while being assessed individually (Feden and Vogel 2003). A common model of cooperative learning found in the literature is the one due to Johnson, and Smith (Johnson et al. 1998a, b). It incorporates five specific tenets, which are *individual accountability*, *mutual interdependence*, *face-to-face promotive interaction*, *appropriate practice of interpersonal skills*, and *regular self-assessment of team functioning*. While different cooperative learning models exist (e.g., Slavin 1983), the core element held in common is a focus on cooperative incentives rather than competition to promote learning.

Problem- and *Project-based learning* are instructional methods where relevant problems and/or complete projects are introduced at the beginning of the instruction cycle and used to provide the context and motivation for the learning that follows. The core element of Problem- and Project-based learning is the significant in context learning that can strongly enhance the student conceptual understanding and motivation.

Flipped classroom is a blended learning approach based on the access to learning resources before attending class, followed by face-to-face sessions that become more and more student-centered via discussion, collaborative learning, and problem-solving activities. This is today enhanced by the increasing availability of digital technology to create the learning resources and easily access them. The core elements are here the possibility for students to access the learning resources before attending class, to move activities, including those that may have traditionally been considered homework, into the classroom, and the possibility to engage in concepts in the classroom with the guidance of a mentor or facilitator.

Inquiry-based learning is an approach based on the possibility for students to ask their questions, to plan research, to collect data from different sources, to construct models, and to elaborate and share results and argumentations. The core elements are here the opportunity for students to act like researchers and the significant amount of self-directed learning on the part of the students.

2.4 Is Active Learning Effective?

The empirical support for AL effectiveness is extensive; however, not all of this support is compelling. Prince (2004) summarizes the literature and discusses the advantages and disadvantages of many types of AL. In the following, we will briefly point out the evidence for the effectiveness of different kinds of AL in helping students to build and develop their knowledge and skills. We will also highlight some issues arising with AL activities that are discussed in the literature.

Active lecturing. Di Vesta and Smith (1979) and Ruhl et al. (1987) report a significant improvement in short-term and long-term retention of relevant concepts in students involved in this kind of active approach to lectures. Moreover, considerable increases in conceptual knowledge building (Streveler et al. 2008) and understanding and modification of common-sense conceptions (Redish et al. 1997; Laws et al. 1999; Reed-Rhoads et al. 2007) are highlighted. However, the traditional lecture environment can be more comfortable for auditory- and reading-type learners, and some students expect and prefer to be passive and have all the answers come from the instructor (Petersen and Gorman 2014).

Collaborative learning. Many studies highlight significant improvement in academic achievement, concept retention, conceptual knowledge, and self-esteem (Johnson et al. 1998a, b; Springer et al. 1999; Sokoloff et al. 2007; Laws et al. 2015). Also, a substantial reduction in attrition rates between students is reported (Berry 1991; Fredericksen 1998).

Cooperative learning. Many studies highlight considerable improvement in academic achievement, concept retention, conceptual knowledge, self-esteem (Sokoloff et al. 2007; Laws et al. 2015; Johnson and Johnson 1989), and promotion of teamwork and interpersonal skills (Laws et al. 2015; Johnson and Johnson 1989; Terenzini et al. 2001; Panitz 1999).

Problem and Project-based learning. Vernon and Blake (1993), Shin and Kim (2013), and Demirel and Dağyar (2016) give evidence of improvement in students' positive attitudes and opinions concerning their programs. Moreover, other studies support the usefulness of these approaches to promote long-term retention of knowledge (Norman and Schmidt 1993; Gallagher 1997; Dochy et al. 2003; Strobel and van Barneveld 2009; Yew and Goh 2016) and an increase in library use, textbook reading, and class attendance (Gallagher 1997; Major and Palmer 2001). Oja (2011) also gives evidence of improvement in critical reasoning skills. However, in some cases, lower scores in short-term retention, in self-efficacy perception, and in traditional evaluation tests have been highlighted (Dochy et al. 2003; Major and Palmer 2001; Oja 2011; Albanese and Mitchell 1993; Gormally et al. 2009; Pourshanazari et al. 2013).

Flipped classroom. Studies like the one due to Morton and Colbert-Getz (2017) highlight how a flipped classroom approach to learning can support higher attainment on questions that required analysis, but no clear difference with more traditional approaches on overall performance is demonstrated. Furthermore, Cheng et al. (2017) highlighted how providing medical students with histology video lectures and quizzes before in-class activities supported greater learning gains, compared to a traditional classroom arrangement. However, despite the positive perceptions of flipped classroom activities, the effects on changes in knowledge and skills are less conclusive and suggest a lack of evidence on its effectiveness (Chen et al. 2017).

Inquiry-based learning. Gormally et al. (2009) report an improvement in student understanding of science processes due to Inquiry-based activities, and Duran and Dokme (2016) give evidence of improvement in critical reasoning skills. Also improvement in conceptual understanding and problem-solving skills has been reported (Lindsey et al. 2012; Persano Adorno et al. 2015), and improvement in critical thinking and the understanding of the nature of science by Yen and Huang (2001), Krystyniak and Heikkinen (2007), and Capps and Crawford (2013). Moreover, Inquiry-based learning has proved to be effective in improving "procedural understanding," critical and reflective thinking (Zion et al. 2004; Sadeh and Zion 2009), in repairing misconception (Prince et al. 2012), in favoring the development of habits of mind conductive to scientific research (Karelina and Etkina 2007), and in strengthening students' practical and reasoning abilities, by proficiently applying the learned concepts to face and solve real-world problem situations (Pizzolato et al. 2014). However, some studies (Trautmann et al. 2004; Quintana et al. 2005) highlighted some undesirable effects of Inquiry learning, such as feelings of inadequacy or frustration in students exposed to Open Inquiry activities (activities sometimes difficult for individuals exhibiting an introverted personality), and a not adequate understanding of concepts. Moreover, difficulty in using Inquiry-based approaches to develop new scientific concepts has been, in some cases, described (Millar 2012).

3 An Example of the Application of Active Learning Strategies

In this example, we will briefly resume the results of a study conducted by the Physics Education Research Group of the University of Palermo, Italy, about the effectiveness of a learning environment based on the Inquiry-Based Science Education (IBSE) approach in improving undergraduate students' lines of reasoning, redirecting them to explicative-like ones. This study is relevant to our discussion about active learning also because the learning activities developed with the students have been planned in the view of the ideas of Growth Mindset and Deliberate Practice discussed above.

3.1 The Research

The study follows an experimental design and is centered on the comparison of the effectiveness of two different learning environments. In them, undergraduate students are involved in the construction and use of explanations of thermally activated phenomena in a context-oriented to the development of a unifying approach to various natural phenomena. In particular, we focused on physical and chemical systems that can exist in two different states characterized by an energy difference ΔE where the state transition is thermally activated by overcoming the potential barrier ΔE. The behavior of these systems is described by expressions all containing the Boltzmann factor. The main aim of this study is to investigate the differences in the efficacy that an Inquiry-based approach focused on the well-known Feynman's unifying approach (FUA) to thermally activated phenomena (Feynman et al. 1963) can have in developing explanation and generalization skills in undergraduate students concerning a more traditional approach, still based on laboratory and modeling activities focused on FUA, but not explicitly developed by using IB teaching/learning methods (Battaglia et al. 2017).

Seventy-two students attending the Undergraduate Program for Chemical Engineering at the University of Palermo, Italy, during Academic Year 2014–2015, were involved in this research. During the first semester of their degree program, the students attended general mathematics, physics, and inorganic chemistry courses and passed the exams. When selected to participate in our study, they were attending a second-semester Physics course dealing with the fundamentals of electromagnetism.

The sample was randomly subdivided into two equally populated groups, an "experimental" group, and a "control" group. The 36 students of the experimental group attended a 20-hour, FUA- and Inquiry-based workshop designing and carrying out their investigations, gathering information, collecting and analyzing data, providing explanations, and sharing their results. The control group students attended a more traditional 20-hour workshop, still based on laboratory and modeling activities based on FUA, but not explicitly developed by following an IB approach. A questionnaire containing six open-ended questions on thermally activated phenomena was administered to the students of both groups before instruction to assess the initial student skills with respect to the explication of natural phenomena. A second one, conceptually similar but related to different physical content, was administered after instruction to study the effects of the two workshops on students' explicative skills.

A quantitative study of the questionnaire responses was done by using cluster analysis techniques (Everitt et al. 2011) aimed at allowing the researcher to group the students in similar subgroups (clusters) and at easily evidencing common patterns in the student responses to the questions (Fazio and Spagnolo 2008; Di Paola et al. 2016; Battaglia et al. 2019). This procedure can help the researcher to infer student lines of reasoning related to the creation and use of explanations in an unsupervised analysis (e.g., Sathya and Abraham 2013).

3.2 The Questionnaires

The reasoning deployed by the students when asked to explain phenomena, and to relate them to the physics and chemistry, they had already studied in previous courses that was analyzed before instruction by using a specially designed and previously validated six-item questionnaire (Fazio et al. 2012, 2013), shown below:

1. A puddle dries more slowly at 20 °C than at 40 °C. Assuming all other conditions (except temperature) equal in the two cases, explain the phenomenon, pointing out what are the quantities needed for the description of the phenomenon and for the construction of an interpretative model of the phenomenon itself.
2. In chemical kinetics, it is well known that the rate of a reaction, u, between two reactants follows the Arrhenius law:

$$u = Ae^{-\frac{E}{kT}}$$

Describe each listed quantity, clarifying its physical meaning and the relations with the other quantities.

3. What do you think the role of a catalyst is in the development of a chemical reaction?
4. Can you give a microscopic interpretation of the Arrhenius law?
5. Can you think of other natural phenomena that can be explained by a similar model?
6. Which similarities can be identified in the previous phenomena? Is it possible to find a common physical quantity that characterizes all the systems you discussed in the earlier questions?

The students in the experimental group then took a 20-hour workshop based on a Bounded/Open Inquiry-based approach and focused on FUA. The workshop dealt with physical content (electricity) different from the one addressed by the questionnaire, but strictly related to the framework of thermally activated phenomena. The students in the control group took a course of equal duration and with the same instructors of the experimental one. During this course, the same physical content and the same FUA approach were dealt with, but the pedagogical methods used were more traditional, still based on laboratory and modeling activities but not focused on Inquiry.

At the end of the workshops, a new questionnaire, validated in the same way of the pre-instruction one, was administered to the students of both groups. It was again aimed at exploring student lines of reasoning about the use of explanations in science. This questionnaire was conceptually similar to the pre-instruction one but was focused on physical/chemical contents (fluidity) not explicitly discussed before and/or during the workshop. It is reported below. All the 36 students in each group completed the post-instruction questionnaire.

1. In modern oil mills, olive oil flows inside metallic pipes. These pipes are often enclosed in larger, coaxial tubes in which hot water flows. Explain the possible reason of this, pointing out what are the quantities needed for a description of the proposed situation and for the construction of an explicative model.
2. In chemistry, it is well known from Eyring's absolute rate theory that fluid viscosity follows the following law:

$$\eta = Ae^{E_{vis}/kT}$$

Describe each listed quantity, clarifying its physical meaning and the relations with the other quantities.

3. In the petroleum industry, additives are often added to gas oil to work as catalysts. What do you think can the role of these additives be in the flowing of gas oil in a pipe?
4. Can you give a microscopic interpretation of the $\eta(T)$ law seen in question 2)?

5. Can you think of other natural phenomena that can be explained by a similar model?
6. Which similarities can be identified in the previous phenomena? Is it possible to find a common physical quantity that characterizes all the systems you discussed in the earlier questions?

3.3 Specific Content and Workshop Methodology

The content of the two workshops dealt with the study of electric current in materials (conductors and semiconductors) and in vacuum systems (thermionic tubes). In particular, situations, where the Boltzmann factor (BF) can be used to describe electric conduction, were analyzed (Battaglia et al. 2010).

3.3.1 Inquiry-Based Workshop

The workshop attended by the experimental group was based on a mixed Bounded/Open Inquiry approach developed through specific Deliberate Practice aimed at the development of student Growth Mindset. It was organized in a series of meetings for a total of 20 h, during which students explicitly followed the 5E phases typical of the IBSE approach (Bybee 1993; Bybee et al. 2006). They could pose their questions and search for sources of information to obtain a solution, in many cases, even proposing and conducting possible experiments and simulations. Building concept maps in each phase of work (i.e., Deliberate Practice) and sharing and contrasting the obtained results in great group discussions were also strongly suggested activities. The students had already studied electric conduction during the lessons of the electromagnetism course. The activities performed during the workshop are briefly resumes as follows:

In the beginning, the students were engaged in the project activities by a discussion about conduction in ohmic conductors and search for evidence of non-ohmic behavior, as in semiconductor devices. Then, students acquired information and planned their activities in small groups, trying to pose questions they would answer during the experimental activities. They were introduced to the laboratory and encouraged to explore the measurement facilities and materials available, to design their own experiences. Students chose to address the electrical conduction process in vacuum tubes, which is easier to discuss and shows marked non-ohmic behavior.

Students carried out their research investigations, designed based on the hypotheses and questions formulated during the explorative phase. They decided to study the anodic current versus the filament temperature, to collect information about the values of concentration of electrons emerging from the filament. Mathematical modeling procedures were discussed to find a law to describe the concentration versus temperature trend, which was found to contain the general BF expression. Some students searched for suitable models to make sense of their experimental evidence and the specific form of the suitable function they found, in particular concerning the meaning of the quantity "energy" contained in the law's exponential term. In manuals and on the Internet, they found references to Richardson's law (Pauling 1988) in vacuum tubes, which is described by an expression analogous to the mathematical function best fitting their experimental data. It contains the BF and students found that the "energy" reported in Richardson's law's exponential term is called the "work function," something conceptually identical to the activation energy. The instructor also suggested analyzing the energy band model in semiconductors and the energy gap concept, by comparing it with the activation energy and work function concepts, discussed before. After a group discussion, the instructor encouraged students to focus on the idea of a "two-level" system as the unifying concept behind all the situations.

Students spent time in the analysis of an agent-based computer model (Battaglia et al. 2009) related to the subject, built by using the NetLogo simulation environment (http://ccl.northwestern.edu/netlogo/), which is very easy to learn, at least in its basic aspects, and can simulate the interactions between a large number of elements. This choice was driven by educational research results that show how mathematical modeling environments, also based on information technologies (Berry et al. 1986; Tarantino et al. 2010), can supply effective pedagogical strategies dealing with complex real-world systems and everyday problem solutions and can help students to develop critical cognitive skills. Students discussed a simulated mechanical model of a two-level system with the instructor. Particularly, they dealt with a large number of balls free to move on two connected planes, placed at different heights. Using a pre-coded, basic NetLogo simulation and then improving the code by building more complex simulations (i.e., Deliberate Practice), it was possible to study the equilibrium distribution of the balls at the two levels and discuss the factors that influence this distribution.

Finally, students compared the simulation findings, the experimental results, and the models explaining them. Students searched for physical and chemical situations different from the ones discussed during the previous activities, whose experimental dependence on temperature gives evidence of similarity with electrical conduction in semiconductors and thermionic tubes. A final scientific report was written by each group, with students sharing their ideas and preliminary results with the other participants. Students presented the most significant findings obtained as a result of their experimental work and held a class discussion aimed at comparing and contrasting the results obtained by different groups. This was considered by the students a crucial stage of the workshop. They maintained that the work of every single group was fundamental to build a shared, final model of the experimental situations they explored. Moreover, they acknowledged that group work and the final in-class discussions greatly helped them to improve their understanding of the subject and their experimental and modeling skills (Growth Mindset).

3.3.2 Traditional Workshop

The same content was also developed during the workshop attended by the control group students. However, the general approach was more traditional in the sense that the students often had to investigate teacher-presented situations and questions through procedures presented by the teacher as the most effective and adequate for the specific aim. Particularly, the students often received detailed instructions on the sources of information to use to make sense of the proposed problems and on the experimental/modeling activities to perform during the various stages of the workshop, basically leading to correct, but predetermined discovery. The students were led to deal with electric conduction process in vacuum tubes and were allowed to use worksheets to take note of their experimental results and report their comments and modeling results. They also used the same simulation tools used by the experimental group students. However, they were never requested to follow the 5E phases typical of IBSE or to present their most significant findings to classmates explicitly. Each student group was asked to write a scientific report of the results obtained during the workshop activities. A final class discussion, mainly lead by the instructor, but still involving students in answering questions and issues proposed by the instructor and aimed at comparing and contrasting the obtained results, was held during the last lesson before the administration of the final questionnaire.

3.4 Analysis of the Answers to Pre- and Post-instruction Questionnaires

The study of student answers was performed by using a specific not-hierarchical clustering method, known as *k-means* (MacQueen 1967), and we will discuss here only the main results of the analysis. More detail on the clustering method, on the number and specific characteristics of clusters of students obtained in the analysis of the answers to both the pre- and post-instruction questionnaires, and on the typical answers given by the students to the questions can be found in Battaglia et al. (2017).

The pre-instruction test results show that both the students that attended the Inquiry-based workshop (experimental group) and the ones that attended a more traditional one (control group) initially highlighted reasoning lines in many cases not well fitted to the study of physics. In fact, during the pre-test, students of both experimental and control group often implemented answering strategies which are inefficient to correctly find a microscopic functioning mechanism of the proposed phenomenon (a drying puddle) and to build proper explanations based on the variables considered relevant for the phenomenon. Only in some cases, the drying of the puddle was explained in terms of a rough functioning mechanism. Very often, students in both groups simply made reference to already known mathematical models. They highlighted a tendency to search in memory for real-life examples or studied concepts that can fit in with the formulas, in some cases without a clear understanding of their physical meaning. Arrhenius law was often described in mathematical form, without a reference to its physical meaning. Finally, in many cases, students highlighted a lack of generalization skills, being limited in their answers to questions 5 and 6 to the context of studied subjects. However, in some cases, a search for a common microscopic model for the situations recalled in answers to the last two questions is present. Resuming, the students of both groups showed in their answers to the pre-instruction questionnaire a significant use of approaches based on common-sense knowledge, even if in some cases in conjunction with descriptive strategies based on the previous study or with a search for rough functioning mechanisms.

The results of the quantitative analysis of student answer to the post-instruction test showed; on the other hand, that a difference between the experimental and the control group students could be identified. Six of the 36 students in the experimental group, placed in the same cluster, were able to explain the situations and problems proposed in the questionnaire relating them to a functioning mechanism based on the idea of thermal activation of molecules. Many of the students included in two other clusters, although in some cases still anchored to memories of past studies, showed to be able to at least explain the flow process in mathematical terms or by citing a functioning mechanism. They discussed the role of additives by considering the energy gap concept, but in some cases, did not relate it to the interaction between molecules. However, in some cases, the Arrhenius-like expression for viscosity was interpreted in terms of the interaction between molecules. Finally, they seemed to possess generalization skills, even if, in some cases, still limited to familiar contexts.

An analysis of student worksheets, of their final reports, and of some interviews taken with a representative student for each cluster confirms that a general shift from descriptive-type reasoning strategies to higher-level ones, based on a search for explanations of the analyzed phenomena, can be highlighted in the experimental group students. Many of the students of this group developed lines of reasoning about the Arrhenius-type phenomena that helped them to build explanations coherent with those of the accepted physical model and correct predictions of the behavior of proposed situations perceived as similar. Moreover, the recognition of the common mathematical form in Arrhenius-like laws was, in many cases, linked to a better understanding of the functioning mechanisms behind these laws, something that was present at a substantially lower level in the initial phases of the workshop. In many cases, before the workshop activities, students focused their reasoning on mathematical descriptions. After the Inquiry-based workshop activities, where they were encouraged to deliberately apply the practice to search for answers to proposed situations and phenomena and to perform measurements and build models in a peer-to-peer setup, many students appeared more confident in looking for microscopic models that can explain the experimental evidence first and then to discuss and make sense of the mathematical law common form. From many reports, it also appeared that the deliberate practice with the two-level system proposed by the simulation supplied students with the support they needed to give meaning to the physical quantities involved in the different phenomena, also stimulating them toward appropriate generalizations.

Some of the control group students also showed a general improvement in reasoning with respect to the one highlighted in the pre-instruction results. In fact, students in one of the post-instruction clusters were able to correctly find and physically interpret the variables relevant in Arrhenius law, to discuss the role of additives in terms of the energy gap concept (although only at the macroscopic level), to give an explanation of the flow process in terms of molecular interaction, and to find and discuss phenomena that can be considered similar to the proposed one. On the other hand, the majority of the other students still based their approaches on reasoning forms based on the memory of studied subjects or on macroscopic or mathematical explanation, without clear evidence of the search for a microscopic functioning mechanism.

Again, an analysis of control group student worksheets, of their final reports, and interviews taken with a representative student for each cluster seemed to confirm these considerations. The main approach to the analysis of the situations proposed during the workshop was anchored to a recall of studied subjects, to the main use of mathematics to give "explanations" (that were rather simply descriptions) and to extensive use of analogies to real-life situations to make sense and explain the new proposed situation. In some cases, this use of analogies was pushed too far and led the students to generalize contents and models wrongly.

4 Conclusion

Active learning methods and strategies allow the students to become leading actors in the development of their learning processes and are credited to be an important means for the development of student critical cognitive skills. However, some teachers and faculty remain skeptical about its real efficacy. Many also express doubts about what AL actually is and how it can be considered different from traditional education and/or do not always understand how the most common forms of active learning differ from each other.

In this paper, after the first definition of AL, we discussed some of the pedagogical and psychological theories AL is based on. We distinguished among different types of AL strategies most frequently discussed in the PER literature and identified basic core elements for each of these types to differentiate among them. Then, we presented a brief review of the literature regarding AL effectiveness in improving student skills and conceptual understanding. Finally, we discussed an example study regarding the effectiveness of an active learning approach developed at the University of Palermo, Italy. We showed that such a learning approach, based on the idea of inquiry and discovery and focused on Deliberate Practice to foster the development of Growth Mindset, can help the students to build mechanisms of functioning and explicative models, and to identify common aspects in apparently different phenomena that at an expert level are all described by the same model.

References

Albanese M, Mitchell S (1993) Problem-based learning: a review of literature on its outcomes and implementation issues. Acad Med 68(1):58–81

Atkinson JW, Feather NT (1966) A theory of achievement motivation. Robert E. Krieger Publishing Company, Huntington

Bandura A (1977) Social learning theory. Prentice Hall, Englewood Cliffs

Barbe WB, Swassing RH, Milone MN (1979) Teaching through modality strengths: concepts practices. Zaner-Bloser, Columbus

Battaglia OR, Bonura A, Sperandeo-Mineo RM (2009) A pedagogical approach to the Boltzmann factor through experiments and simulations. Eur J Phys 30:1025–1037

Battaglia OR, Fazio C, Guastella I, Sperandeo-Mineo RM (2010) An experiment on the velocity distribution of thermionic electrons. Am J Phys 78(12):1302–1308

Battaglia OR, Di Paola B, Persano Adorno D, Pizzolato N, Fazio C (2017) Evaluating the effectiveness of modelling-oriented workshops for engineering undergraduates in the field of thermally activated phenomena. Res Sci Educ. https://doi.org/10.1007/s11165-017-9660-0

Battaglia OR, Di Paola B, Fazio C (2019) Unsupervised quantitative methods to analyze student reasoning lines: theoretical aspects and examples. Phys Rev Phys Educ Res 15(2):020112

Berry L Jr (1991) Collaborative learning: a program for improving the retention of minority students, ERIC # ED384323

Berry JS, Burghes DN, Huntley ID, James DJG, Moscardini AO (1986) Mathematical modelling. Methodology, models and micros. Wiley, New York

Bonwell CC, Eison JA (1991) Active Learning: creating excitement in the Classroom. ASHEERIC Higher Education Report no 1. George Washington University, Washington, DC

Bybee RW (1993) An instructional model for science education. Developing biological literacy. Biological Sciences Curriculum Study, Colorado Springs

Bybee RW, Taylor JA, Gardner A, Van Scotter P, Carlson Powell J, Westbrook A, Landes N (2006) The BSCS 5E instructional model: origins and effectiveness. Biological Sciences Curriculum Study, Colorado Springs

Capps DK, Crawford BA (2013) Inquiry-based instruction and teaching about nature of science: are they happening? J Sci Teach Educ 24(3):497–526

Chen F, Lui MA, Martinelli SM (2017) A systematic review of the effectiveness of flipped classrooms in medical education. Med Educ 51:585–597

Cheng X, Ka Ho Lee K, Chang EY, Yang X (2017) The 'flipped classroom' approach: stimulating positive learning attitudes and improving mastery of histology among medical students. Anat Sci Educ 10:317–327

Cummings K (2013) A community-based report of the developmental history of PER. Paper presented at the American Association of Physics Teachers, Portland, Oregon

Demirel M, Dağyar M (2016) Effects of problem-based learning on attitude: a meta-analysis study. Eurasia J Math Sci Technol Educ 12(8):2115–2137

Dewey J (1916) Democracy and education. An introduction to the philosophy of education. The Macmillan Company, New York

Di Paola B, Battaglia OR, Fazio C (2016) Non-hierarchical clustering as a method to analyse an open-ended questionnaire on algebraic thinking. S Afr J Educ 36(1):#1142

Di Vesta F, Smith D (1979) The pausing principle: increasing the efficiency of memory for ongoing events. Contemp Educ Psychol 4(3):288–296

Dochy F, Segers M, Van den Bossche P, Gijbels D (2003) Effects of problem-based learning: a meta-analysis. Learn Instr 13:533–568

Duran M, Dokme I (2016) The effect of the inquiry-based learning approach on student's critical thinking skills. Eurasia J Math Sci Technol Educ 12(12):2887–2908

Dweck C (2006) Mindset: the new psychology of success. Random House, New York

Ericsson KA (2004) Deliberate practice and the acquisition and maintenance of expert performance in medicine and related domains. Acad Med 79(10 Suppl):S70–S81

Everitt BS, Landau S, Leese M, Stahl D (2011) Cluster analysis. Wiley, Chichester

Fazio C, Spagnolo F (2008) Conceptions on modelling processes in Italian high-school prospective mathematics and physics teachers. S Afr J Educ 28(4):469–487

Fazio C, Battaglia OR, Guastella I (2012) Two experiments to approach the Boltzmann factor: chemical reaction and viscous flow. Eur J Phys 33(2):359–371

Fazio C, Battaglia OR, Di Paola B (2013) Investigating the quality of mental models deployed by undergraduate engineering students in creating explanations: the case of thermally activated phenomena. Phys Rev ST Phys Educ Res 9:020101

Feden P, Vogel R (2003) Methods of teaching: applying cognitive science to promote student learning. McGraw Hill Higher Education, Boston

Feynman RP, Leighton RB, Sands M (1963) The Feynman lectures on physics, vol I. Addison-Wesley, Reading, pp 42-1–42.11

Fleming ND, Mills C (1992) Not another inventory, rather a catalyst for reflection. Improve Acad 11:137–155

Fredericksen E (1998) Minority students and the learning community experience: a cluster experiment, ERIC # ED216490

Gallagher S (1997) Problem-based learning: where did it comes from, what does it do and where is it going? J Educ Gifted 20(4):332–362

Georgiou H, Sharma MD (2015) Does using active learning in thermodynamics lectures improve students' conceptual understanding and learning experiences? Eur J Phys 36:015020

Gormally C, Brickman P, Hallar B, Armstrong N (2009) Effects of inquiry-based learning on students' science literacy skills and confidence. Int J Sch Teach Learn 3(2):Article 16

Hake RR (1998) Interactive-engagement versus traditional methods: a six-thousand-student survey of mechanics test data for introductory physics courses. Am J Phys 66:64–74

Herrington J, Oliver R (2000) An instructional design framework for authentic learning environments. Educ Technol Res Dev 48(3):23–48

Johnson D, Johnson R (1989) Cooperation and competition, theory and research. Interaction Book Company, Edina

Johnson D, Johnson R, Smith K (1998a) Active learning: cooperation in the college classroom, 2nd edn. Interaction Book Co., Edina

Johnson D, Johnson R, Smith K (1998b) Cooperative learning returns to college: what evidence is there that it works? Change 30(4):26–35

Jung CG (1971) Psychological types. Princeton University Press, Princeton

Karelina A, Etkina E (2007) Acting like a physicist: student approach study to experimental design. Phys Rev Spec Top Phys Educ Res 3:020106

Krystyniak RA, Heikkinen HW (2007) Analysis of verbal interactions during an extended, open-inquiry general chemistry laboratory investigation. J Res Sci Teach 44:1160

Laws P, Sokoloff D, Thornton R (1999) Promoting active learning using the results of physics education research. UniServe Sci News 13:14–19

Laws PW, Willis MC, Sokoloff DR (2015) Workshop physics and related curricula: a 25-year history of collaborative learning enhanced by computer tools for observation and analysis. Phys Teach 53(7):401–406

Lindsey BA, Hsu L, Sadaghiani H, Taylor JW, Cummings K (2012) Positive attitudinal shifts with the Physics by Inquiry Curriculum across multiple implementations. Phys Rev ST Phys Educ Res 8:010102

MacQueen J (1967) Some methods for classification and analysis of multivariate observations. In: LeCam LM, Neyman J (eds) Berkeley symposium on mathematical statistics and probability 1965/66, vol I. Univ. of California Press, Berkeley, pp 281–297

Major C, Palmer B (2001) Assessing the effectiveness of problem-based learning in higher education: lessons from the literature. Acad Exch Q 5(1):4

Mayer RE (2008) Learning and instruction. Pearson Education, Inc., Upper Saddle River

Millar R (2012) Rethinking science education: Meeting the challenge of "science for all". Sch Sci Rev 93:21

Morton DA, Colbert-Getz JM (2017) Measuring the impact of the flipped anatomy classroom: the importance of categorizing an assessment by Bloom's taxonomy. Anat Sci Educ 10:170–175

Myers IB, McCaulley MH, Quenk NL, Hammer AL (1998) MBTI manual: a guide to the development and use of The Myers-Briggs Type Indicator, 3rd edn. Consulting Psychologists Press, Palo Alto

Norman G, Schmidt H (1993) The psychological basis of problem-based learning: a review of evidence. Acad Med 67:557–565

Oja KJ (2011) Using problem-based learning in the clinical setting to improve nursing students' critical thinking: an evidence review. J Nurs Educ 50(3):145–151

Panitz T (1999) The case for student centered instruction via collaborative learning paradigms, ERIC # ED 448444

Pauling L (1988) General chemistry. Dover, New York, p 551

Pelley J (2014a) Making active learning effective. Med Sci Educ 24(Suppl 1):S13–S18

Pelley JW (2014) Learning style: implications for teaching and learning. In: Matheson (ed) An introduction to the study of education, 4th edn. Routledge, London

Pelley JW, Dalley BK (2008) Success types in medical education. Retrieved 21 Dec 2018 from: https://www.ttuhsc.edu/medicine/medical-education/success-types/documents/stsinmeded.pdf

Persano Adorno D, Pizzolato N, Fazio C (2015) Elucidating the electron transport in semiconductors via Monte Carlo simulations: an inquiry-driven learning path for engineering undergraduates. Eur J Phys 36(5):055017

Petersen C, Gorman K (2014) Strategies to address common challenges when teaching in an active learning classroom. New Dir Teach Learn 137:63–70

Pizzolato N, Fazio C, Sperandeo-Mineo RM, Persano-Adorno D (2014) Open-inquiry driven overcoming of epistemological difficulties in engineering undergraduates: a case study in the context of thermal science. Phys Rev Spec Top Phys Educ Res 10:010107

Pourshanazari A, Roohbakhsh A, Khazaei M, Tajadini H (2013) Comparing the long-term retention of a physiology course for medical students with the traditional and problem-based learning. Adv Health Sci Educ 18(1):91–97

Prince M (2004) Does active learning work? A review of the research. J Eng Educ 93(3):223–231

Prince M, Vigeant M, Nottis K (2012) Using inquiry-based activities to repair student misconceptions related to heat, energy and temperature. In: Frontiers in education conference proceedings (FIE), Seattle, WA, pp 1–5. https://doi.org/10.1109/fie.2012.6462344

Pritchard A (2009) Ways of learning: Learning theories and learning styles in the classroom, 2nd edn. Routledge, New York

Quintana C, Zhang X, Krajcik J (2005) A framework for supporting meta cognitive aspects of online inquiry through software-based scaffolding. Educ Psychol 40:235

Redish EF, Smith KA (2008) Looking beyond content: skill development for engineers. J Eng Educ 97(3)

Redish E, Saul J, Steinberg R (1997) On the effectiveness of active-engagement microcomputer-based laboratories. Am J Phys 65(1):45–54

Reed-Rhoads T, Imbrie P K, Allen K, Froyd J, Martin J, Miller R, Steif P, Stone A, Terry R (2007) Tools to facilitate better teaching and learning: concept inventories. Panel at ASEE/IEEE Frontiers in Education Conference, Milwaukee, WI

Revans RW (1982) The origin and growth of action learning. Chartwell-Bratt, Brickley

Ruhl K, Hughes C, Schloss P (1987) Using the pause procedure to enhance lecture recall. Teach Educ Spec Educ 10:14–18

Sadeh I, Zion M (2009) The development of dynamic inquiry performances within an open inquiry setting: a comparison to guided inquiry setting. J Res Sci Teach 46:1137

Sathya R, Abraham A (2013) Comparison of supervised and unsupervised learning algorithms for pattern classification. Int J Adv Res Artif Intell 2(2):34–38

Sharma MD, Johnston ID, Johnston HM, Varvell KE, Robertson G, Hopkins AM, Thornton R (2010) Use of interactive lecture demonstrations: a ten year study. Phys Rev Spec Top-Phys Educ Res 6:020119

Shin IS, Kim JH (2013) The effect of problem-based learning in nursing education: a meta-analysis. Adv Health Sci Educ Theory Pract 18(5):1103–1120

Slavin R (1983) Cooperative learning. Research on teaching monograph series, ERIC digest ED242707

Smith B, MacGregor J (1992) What is collaborative learning? In: Goodsell A et al (eds) Collaborative learning: a sourcebook for higher education. National Center on Postsecondary Teaching, Learning and Assessment, University Park, pp 9–22

Sokoloff DR, Thornton RK, Laws PW (2007) RealTime physics: active learning labs transforming the introductory laboratory. Eur J Phys 28:S83–S94

Springer L, Stanne M, Donovan S (1999) Effects of small-group learning on undergraduates in science, mathematics, engineering and technology: a meta-analysis. Rev Educ Res 69(1):21–52

Streveler RA, Litzinger TA, Miller RL, Steif PS (2008) Learning conceptual knowledge in the engineering sciences: overview and future research directions. J Eng Educ 97(3):279–294

Strobel J, van Barneveld A (2009) When is PBL more effective? A meta-synthesis of meta-analyses comparing PBL to conventional classrooms. Interdiscip J Probl-based Learn 3(1):44–58

Tarantino G, Fazio C, Sperandeo-Mineo RM (2010) A pedagogical flight simulator for longitudinal airplane flight. Comput Appl Eng Educ 18(1):144–156

Terenzini P, Cabrera A, Colbeck C, Parente J, Bjorklund S (2001) Collaborative learning vs. lecture/discussion: students' reported learning gains. J Eng Educ 90(1):123–130

Trautmann N, MaKinster J, Avery L (2004) What makes inquiry so hard? (and why is it worth it). In: Proceedings of the annual meeting of the National Association for Research in Science Teaching, Vancouver, BC, Canada. http://ei.cornell.edu/pubs/NARST_04_CSIP.pdf

Vernon D, Blake R (1993) Does problem-based learning work? A meta-analysis of evaluative research. Acad Med 68(7):550–563

Vygotsky LS (1986) Thought and language (trans: Kozulin A). The MIT Press, Cambridge, MA

Yen C, Huang S (2001) Authentic learning about tree frogs by preservice biology teachers in open-inquiry research settings. In: Proceedings of the National Science Council Republic of China, ROC(D), vol 11. Taiwan National Science Council, Taipei, p 1

Yew EHJ, Goh K (2016) Problem-based learning: an overview of its process and impact on learning. Health Prof Educ 2:75–79

Zion M, Slezak M, Shapira D, Link E, Bashan N, Brumer M, Orian T, Nussinowitz R, Court D, Agrest B, Mendelovici R, Valanides N (2004) Dynamic, open inquiry in biology learning. Sci Educ 88:728

Dialogue on Primary, Secondary and University Pre-service Teacher Education in Physics

Marisa Michelini

Abstract Teacher education is one of the main fields in Physics Education Research. In the scenario of GIREP contributions in that field, the special case of Italy is presented with the quality of primary teacher education and the difficulties in secondary teacher education, which is then discussed here on the light of the main researches and EU projects. The analysis of the 115 answers to a questionnaire evidence the main needs in teacher education of researchers and teachers of the different school levels.

1 Introduction

Teacher education in physics (TEiP) is a problem having a different nature in different levels of instruction from kindergarten to university. Research results from Physics Education Research (PER) at international level and from European Projects underline the crucial role of TEiP in enhancing the quality of physics education (Sassi and Michelini 2014a).

The PISA, ROSE and TIMMS projects[1] developed case studies showing evidence of the impact by teacher's education as a qualitative improvement in the physics learning environment.

---

[1]PISA 2009: 34 OECD members + 41 partner countries, *PISA 2009 Results: Executive Summary;* (OECD Programm for International Student Assessment—PISA PISA www.pisa.oecd.org/ every 3 years 15 years students?? Assessed in Reading, Mathematical and Scientific literacy; ROSE The Relevance of Science Education ROSE http://www.uv.uio.no/ils/english/research/projects/rose/; TIMSS Trends in International Mathematics and Science Study http://www.timss.bc.edu/;TIMSS 2007: 59 countries, 6 benchmark participants; 4th and 8th grades; about 434,000 students; 47,000 teachers, 15,000 school principals. See Sassi and Michelini 2014b.

M. Michelini (✉)
Physics Education Research Unit, DMIF (Department of Mathematics, Informatics and Physics), University of Udine, Via delle Scienze 206, 33100 Udine, UD, Italy
e-mail: marisa.michelini@uniud.it

© The Editor(s) (if applicable) and The Author(s), under exclusive license to Springer Nature Switzerland AG 2020
J. Guisasola and K. Zuza (eds.), *Research and Innovation in Physics Education: Two Sides of the Same Coin*, Challenges in Physics Education,
https://doi.org/10.1007/978-3-030-51182-1_3

The problem of TEiP is a multidimensional challenge involving the competences needed, the problems encountered up to now, the supports to be provided and the basic requirements of teacher education for primary and secondary schools and for university innovation.

In the past, PER offered studies and results mainly related to physics teachers at secondary level. GIREP in particular has promoted studies on TeiP since its 2000 Conference in Barcelona (Pinto and Surinach 2001).

The meeting of European Ministries of Education in Portugal produced in the same year the *Green Paper on teacher education in Europe* (Buchberger et al. 2000). The Green Paper highlights the crucial role of designing appropriate teaching/learning situations in which prospective teachers can find opportunities to develop the main professional skills as well as a basic scientific culture enabling them to perform successful educational design in spite of their limited knowledge of the subject. The suggested basic activities for teacher education are the following: educational reconstruction of subject matter, problem-solving situations, research-based curricular design, planning teaching/learning interventions, learning knots analysis and students reasoning analysis in T/L activities.

Out of the GIREP Seminar held in Udine in 2003 on *Quality development of teacher education and training* (Michelini 2004a), three main recommendations emerged: (1) *specific professional programmes for teacher education have to be organized in all countries*, (2) didactic research has to be integrated with teaching and teacher education, (3) cooperation between school and university has to be organized for the quality development of teachers. In the Udine Seminar, we find contributions on the problem of primary teacher education at scientific level and in physics especially (Corni et al. 2004; Michelini 2004c) and on research-based intervention modules on this problem (Michelini et al. 2004).

There are several studies on TEiP carried out within the GIREP framework. Hereunder, we are going to present the milestone from the 2000 Barcelona Congress to the Dialogue initiative in San Sebastian (2018). Firstly, we will discuss the special case of the Italian programme for teacher education, as it evinces an interesting approach to the teacher education problem, due to the fact that it takes into account research results in that field, the only advantage produced by the late activation in the year 2000.[2] We will focus here in particular on the recent research-based implementation of primary teacher education in physics in Italy.

2 Italy as a Special Case in Teacher Education and Primary Teacher Education in Physics

Primary teacher education in each scientific subject, as in physics, is particularly important in Italy where a recent reform promoted by Luigi Berlinguer, the Ministry

---

[2]Though activated only in 2000, the need for teacher education was established by law in 1945.

of Education launching the Bologna process for university reform, too, introduced a teacher education programme for both primary and secondary school teachers.

Secondary teacher education in that programme was organized around 2 years of Specialization School post-master degree for each teaching subject. It was organized in the following four areas of the same weight (25–30/120 credits) for teaching qualification: (1) pedagogy, psychology and socio-anthropology, (2) subject education, (3) educational laboratories, (4) apprenticeship. Content knowledge (CK) of the subject was considered a pre-requisite controlled by means of an entrance exam. In each area, prospective teachers were asked to have an exam every 6 credits and a final exam on a teaching practice thesis. This very good project was a training ground for the university itself, not yet prepared for teacher education. A few years later, the political situation changed and the new Ministry of Education transformed the secondary teacher education into a one-year post-master course. Another change in the national policy modified once more the situation, and Italy currently is facing a very bad situation as regards secondary teacher education: we will hopefully be able to discuss a better programme in future.

This is not the case of primary teacher education, which is an excellence up to now. For primary teacher education, Italy has a specific 5-year university degree.[3] The primary teacher degree includes 78 credits (26% of the curriculum for pedagogy, anthropology psychology and sociology, 135 credits of subject education) and educational laboratories (44% of the curriculum). Apprenticeship is jointly organized by school and university (8%) with the support of teachers who are working part time (50%) at university. In recent years, more attention is being lavished on the apprenticeship and on the relationship between laboratories and different subject educations.

Physics Education and Lab is a specific course of 8 + 1 credits: this is a value, but value produces the problem of how to perform that course.

Let us reflect on the scientific education at primary schools and the related teacher education.

Teacher education for scientific primary education is a new challenge that involves the possibility of transferring to future generations a culture in which science is an integral, not a marginal part, not a marginal one; it involves the possibility of equipping students with the fundamental elements of scientific education that allows them to manage them in games, in curious questions, in moments of organized analysis. It is a new challenge compared to the open social goals of primary education in the past. This challenge implies that pupils become aware of what an evidence-based assertion means and how to perform a methodological scientific goal, learning at the same time social behaviour in sharing ideas and discussing hypotheses. Scientific education must offer the opportunity to grow a scientific way of thinking and gradually understand how physics look at phenomena compared to the other sciences, e.g. biology. What this means for pupils is to transform a simple collection of observations into interpretative ideas and develop formal thinking (Michelini 2010). The

---

[3]2000 saw the launch of the 4-year degree, while in 2011, it was transformed into a 5-year program including skills associated with disability.

integrated and interdisciplinary approach in primary teacher education requires a gradual grasp of the identity of subjects and an attention to pupils' ideas to trigger off a conceptual change from common sense ideas to scientific ones (Vosniadou 2008; Amin et al. 2014).

Physics Education Research (PER), in fact, cues us on how pupils steer observation having an interpretative idea, often implicit, and how for scientific learning it is important to help them express ideas and to share interpretative hypotheses.[4] We have to change the teaching style based on the transmission of experiences, in which we have the show of phenomena, the request to observe and to collect observations in a style focusing on active role of students, stimulating pupils to be operative, in the promotion of scientific reasoning and inquiry approach for explanations and interpretations in the context of Conceptual Labs of Operative Exploration (CLOE) (Michelini 2006; Michelini and Stefanel 2016).

While in service, teachers are not prepared for these goals and they cannot count on supporting materials, because there is a lack of qualified teaching resources for primary schools in the scientific field. Textbooks for children are informative and often confuse concepts.

The actual lack of interest in physics is due to the illiteracy of citizens in the scientific field[5]: physics would be more appreciated if pupils started studying it early, avoiding its presentation based purely on facts and rules. If our mission is to increase the relevance of scientific knowledge for all citizens, we have to offer a good pre-service education to primary school teachers. This is a challenge for all European universities (Michelini and Sperandeo Mineo 2014). PER has to support the transition phase from the actual information-based teaching style to an inquiry-based scientific learning style.

Pre-service teacher education has to find a way to offer the new generation of teachers the skills to exploit children's curiosity and spontaneous questions to build scientific thinking in an interactive way educating children to be active and responsible learners.

Relevant problems to solve for such a challenge are: (1) the lack of prospective primary teacher (PPT) skills on the subject (CK). (2) The difficulties novices encounter in putting into practice pedagogical knowledge (PK) in relation to an appropriate CK. (3) The general difficulty to integrate PK and CK for pedagogical content knowledge (PCK) development (Shulman 1987). (4) Skilfulness in the construction of coherent teaching/learning paths (Michelini and Sperandeo Mineo 2014).

The Italian *Physics Education and Lab* course for PPT offers the opportunity to study how to build integrated skills in the subject matter and in pedagogic aspects for the professional didactic skills, particularly those professional skills related to the use of contextual strategies, aimed at helping children overcome conceptual knots and/or activate interpretative models fostering scientific thinking (Elbaz 1983;

---

[4]A large amount of contributions on children's interpretative ideas can be find in the proceedings of Early Science section of ESERA 2015, 2017 conference.

[5]PISA project results are showing a substantial scientific illiteracy at all levels.

Michelini 2001; Borko 2004; Abell 2007; Berger et al. 2008; Davis and Smithey 2009).

The basic choice adopted by us at the Udine Research Unit is the integration of Metacultural, Experiential and Situated training models into the MES model for TEiP (Michelini 2004b; Michelini et al. 2013) and focuses on the construction of a flexible pedagogical content knowledge (PCK), based on Physics Education Research (PER).

Figure 1 shows how in MES model the CK includes the discussion of subject matter and nature of science (NOS), integrated into educational proposals taken from the PER literature and how different proposals on the same topic are discussed in the light of learning difficulties emerging from PER researches. The PK part linked by CK includes the discussion of some crucial aspects, such as the role of operational behaviour or representations or metaphors in scientific learning and the interplay of math and physics, in carrying out an analysis of the proposal from a didactic point of view (rationale, choice of materials, strategies, methods, laboratory activities, … metacultural aspects), living by means of tutorials the same experience of children in the conceptual path steps (experiential aspect) and, after a planning experience, a situated learning experience working with children.

The theoretical framework adopted for the research is the Model of Educational Reconstruction (MER) (Duit et al. 2005). The following five activities characterize path planning and its implementation in class by PPTs: (1) conceptual reconstruction of subject matter, (2) analysis of main conceptual difficulties of specific topic, (3) analysis of research-based educational proposals, (4) reflection on how the main

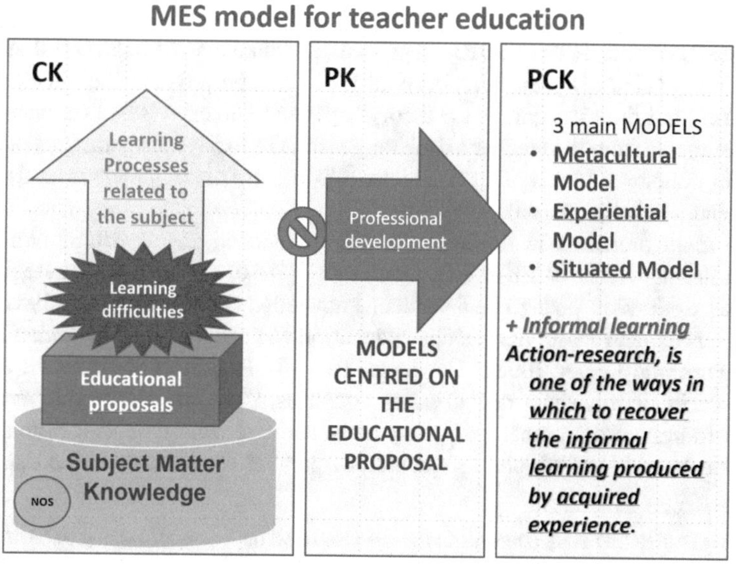

Fig. 1 MES model for teacher education integrating meta-cultural, experiential and situated models in developing PCK

conceptual activities are dealt with in the proposed paths, (5) group discussion about approaches, strategies, activities, instruments and methods as suggested in explored proposals. Inquiry-Based Learning (IBL), Prevision–Experimentation–Comparison (PEC) and Conceptual Labs of Operative Exploration (CLOE) strategies are often discussed (Fedele et al. 2005; Michelini and Stefanel 2015). The following are relevant for PPT education: the personal involvement of PPTs in analysing educational proposals for primary schools, in planning and revising the plan after peer discussion and large group discussion with the person in charge the course and practical classroom implementation of formulated proposals, as evinced by different research-based implementations (Corni et al. 2004; Michelini 2003; Testa and Michelini 2007; Michelini et al. 2011, 2014; Leto and Michelini 2014; Michelini and Mossenta 2014; Vercellati and Michelini 2014; Michelini and Vidic 2016; Vidic et al. 2020).

3   Secondary School Teacher Education in Physics

Eurydice (1998, 2003) offers a rich amount of information on *The teaching profession in Europe*. Several EU funded projects have addressed such main problems of secondary school teacher education from different viewpoints: hands-on experiment, laboratory work, contributions from ICT and ET and informal education, in order to gain possible common frameworks based on experimented examples of good practices. The enquiries conducted have evidenced the need for research work and support for teacher education in physics for professional development of in-service teachers as well as pre-service teacher education.

TIMSS Advanced (2008–2010), in particular,[6] found that the book is still the main educational tool used by teachers; in about 100% of surveyed countries, in more than 50% of school time students read "theory" or how to do exercises. Demonstrations of experiments from the teacher's desk range from 11 to 54%. Experiments or investigations done by students range from 0 to 30% and use of computers from 0 to 50% (Sassi and Michelini 2014a).

The main problems in teacher education evidenced at the GIREP Workshop held in Reims-GIREP-ICPE-MPTL Conference (2010) are related to strategies and methods to develop pedagogical content knowledge (PCK) (Shulman 1987) both in pre-service and in-service teacher education (Michelini and Sperandeo Mineo 2014; Park and Oliver 2008). The Steps-Two EU Project[7] questionnaire results, discussed in Reims, show that in many countries, there are new developments in teacher training programmes and/or methods, and two main models for pre-service teacher education are adopted: sequential or parallel disciplinary and pedagogical

---

[6]TIMSS ADVANCED 2008 (students in the last year of secondary school taking or having taken courses in advanced Mathematics and Physics: Mechanics, E&M, Heat&Temperat., Atoms, Nuclei). Ten countries: AM, IR, IT, LB, NL, NO, PH, RU, SI, SE. Changes tracked in 1995–2008: 5 Countries.

[7]Steps-Two EU Project involved 74 Physics Departments from 32 countries and was supported by EPS. It had a specific Working Group (WG3) on Physics Teacher Education. http://www.stepst wo.eu/.

education; what emerges however from the research results about the nature and level of pre-service physics teacher knowledge is that understanding of subjects delivered during pre-service teacher education courses is not the conceptual understanding that pre-service teacher will need to develop in their future students. The literature on PCK documented different approaches and instruments (tests, video) on measuring teachers' PCK, PCK-in-action and PCK-on-action (reflective component of teachers) and developing a model of professional action-based skills. Some of the main problems emerging are: (a) the lack of skills in conceptual science matter knowledge (SMK); the need for teachers to gain ownership in the concepts and the ways in which physicists interpret phenomena; scientific knowledge and natural reasoning often co-exist within the same terrain; (b) the tendency to reinforce spontaneous ideas and local visions of common sense: the general difficulty to integrate PK and SMK for the construction of PCK; (c) adoption of a transmission style of teaching notions instead of starting from students' ideas to develop their own reasoning; teaching style reproduces the narrative listing of notions: answers to questions not posed; student common sense reasoning is evocated as educational strategy to involve students, but it is not used as a starting point to produce evolution of students' way of thinking, while local to global interpretative perspectives are not promoted; (d) lack of coherence in teaching/learning paths (Michelini and Sperandeo Mineo 2014). Some good practices were discussed in Reims (Michelini 2006; Aiello and Sperandeo Mineo 2001; Sperandeo Mineo et al. 2006; Bozzo et al. 2010; Michelini and Stefanel 2008). Within this framework, the professional preparation of a science teacher has been deeply analysed in terms of *professional profile* in the context of jobs for "Human Talent Management" *skills* (Dineke et al. 2004).

Steps-Two EU Project produced another important contribution to the TEiP with the document titled *European Benchmarks for Physics Teacher Education Degrees*, presented at the World Conference on Physics Education, Istanbul, July 2012 by Urbaan Titulaer[8] (Tasar 2012). According to that document, the Central Requirement for PTE Education should be academics, preferably at master level; research based on the three components: physics, didactic of physics (teaching/learning) and applied pedagogy and social aspects, containing initial practical training in schools, thesis on T/L activities. For example, physics teaching competences are: (1) making clear what science and physics are, promoting scientific literacy and a disposition for inquiry and further learning; (2) offer physics to pupils, using multiple representations and bridging that with pupils' daily experiences; (3) designing a T/L plan within the given constraints; (4) implementing this plan, choosing and designing course material, evaluating their efficacy and learning from experiences made; (5) knowledge of and experience with a broad spectrum of teaching methods, including school experiments and the use of multimedia; (6) identifying students' conceptual difficulties and organizing learning environments for overcoming them.

---

[8]Task force members for the document presented are: *Ovidiu Caltun, Iasi, RO; Eamonn Cunningham, Dublin, IE; Gerrit Kuik and Ed van den Berg, Amsterdam, NE; Marisa Michelini, Udine, IT; Gorazd Planinsic, Ljubljana, SI; Elena Sassi, Napoli, IT; Urbaan Titulaer, Linz, AT (Chair); Rita van Peteghem, Antwerpen, BE; Frank van Steenwijk, Groningen, NL; Vaggelis Vitoratos, Patras, GR.*

Recently, the Hope EU Project published its results (2017),[9] which include an enquiry on teacher education analysed by the Working Group 4. The last activity of the Hope Project was a forum held in Constanta, Romania, on September 2016 on teachers' needs with the participation of representatives of the 71 project partners. In the conclusions, it was underlined that the relevant aspects that need research are: (1) how to test PCK (instruments and methods); (2) how to promote methodological competences related to experimental exploration, modelling and building formal thinking; (3) how to promote argumentation in discourse, description vs interpretation in phenomena exploration and representation role in promoting learning. In the general conclusions, the recommendations sent to the EU Community were: (A) school university cooperation in teacher education and innovation on teaching/learning have to be promoted and supported; (B) PER includes applied research activities and has to be integrated in teacher education and in classroom praxis; (C) ministry of education in EU has to agree on goals and learning outcomes— standards and guidelines, assuming the responsibility to autonomously apply the shared principle in the different contexts. We hope that EU promotes a Task Force on Teacher Education, like those in the USA,[10] to support PER research in that field and to transform into guidelines the PER results and good examples developed up to now.

4 The Dialogue 1 Initiative and Questions on Primary, Secondary and University Pre-service Teacher Education

A lot of work (research and experiences) on teacher education is now available in all countries in the world. We, as GIREP community, have to collect needs and results, promoting research and intervention modules in that field. Inspired by this idea, Jenaro Guisasola, Kristina Zuza and I organized the *Dialogue 1 on Primary, Secondary and University Pre-service Physics Teacher Education* with two preparatory actions. The first action was to prepare and ask six questions in advance on this problem to participants at the San Sebastian Conference, collecting and summarizing results. The second action was to invite experts (Knut NEUMANN,[11] Gabriela LORENZO,[12] Laurence Viennot[13]) coming from very different research approaches to offer along with me an overview of the problem from their perspective, thinking

---

[9]HOPE EU Project is an LLP network project involving 71 Universities. www.hope.org.

[10]National Task Force on Teacher Education in Physics (NTFTEP, USA) http://www.ptec.org/tas kforce.

[11]From Leibniz Institute for Science and Mathematics Education (IPN) at the University of Kiel, Germany.

[12]Universidad de Buenos Aires, Facultad de Farmacia y Bioquímica—Consejo Nacional de Investigaciones Científicas y Técnicas in Argentina.

[13]And Nicolas D´ecamp, Laboratoire de Didactique Andr´e Revuz, EA 4434, Universit´e Paris Diderot, Paris, France.

of the same questions and considering the summarized additional questions posed by participants. By means of this preparation, we offer participants the opportunity to have a real role in a plenary session.

Table 1 sets out the six questions posed by us, while Fig. 2 shows the grouped distribution of the 115 questions posed by participants.

Table 1 Questions posed to participants and to experts in Dialogue 1

1. We have three different teacher profiles: primary teachers, secondary school teachers and university teachers. From your research evidence,
a. which are the main problems for each profile? How do you find solutions?
b. what sort of activities in teacher education promote their professional skills? How?
c. how can the different skills needed by teachers at the different levels considered be integrated?

2. How can prospective teachers find opportunities to develop the main professional skills:
a. which is the role of psychological-pedagogical education?
b. how can the contents for teaching conceptual skills be discussed?
c. Which kind of laboratory proves useful?
d. How to conduct apprenticeships?

3. How can physics teacher be prepared for a significant integration of:
a. ICT in school activities?
b. Lab work in school activities?

4. How to integrate Physics Education Research into physics teacher education?

5. How can physics teaching ability be evaluated?

6. How can we carry out research on teacher education now?

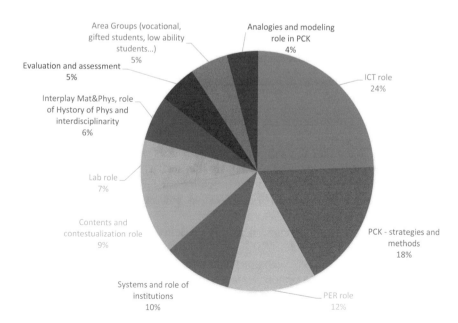

Fig. 2 Grouped distribution of 115 questions posed by participants

Let us exemplify the questions posed.

Some questions are very general, such as the following:

1. Do we really want to develop critical thinking in teachers, and how to proceed in this regard?
2. How does one ensure that a solid physics training stays central to the preparation of a teacher in this technological world?
3. How should a physics teacher take into account a growing multidisciplinary world?
4. If we do not have to improve the scientific education/curriculum because it is good enough, can we be more selective in the qualification process of physics teachers?
5. In Brazil, the best students do not want to become science teachers. What initiatives could reverse this situation in a developing country where the teaching career is so undervalued?
6. In which way could membership of a science educational research team influence professional teacher development?
7. Is it desirable to have educated scientists as teachers in primary and secondary education?
8. What is the role of physics departments in the preparation of physics teachers?
9. What is the situation of university physics teachers' training?
10. What kind of training/knowledge (if any) should a physics teachers' educator have?
11. Which examples of new technology are suited for teacher education?
12. What pedagogical content knowledge does a good physics teacher need?
13. Why is the history of science still more factual than processual and epistemological in teachers' preparation?
14. How can we support teachers to teach concepts of quantum physics?

24% of questions relate to ICT and technologies. Examples of the main questions posed are the following:

1. Would it not be more interesting to put technological knowledge into practice in classroom?
2. Do we need professional development for in-service teachers that deals with the way devices students know from everyday life, such as refrigerators, work?
3. Does university research consider the technological conditions at school?
4. Which new technological techniques would you exploit to teach complex concepts such as quantum mechanics?
5. How big, in your opinion, is the risk to hide a lack in educational meaning behind technology?
6. How can robotics and programming be used for integrated STEM education?
7. How engaging are these activities (related to technology) for physics students?
8. How important is it for teachers to recognize the technological implications of the basic physics they teach?

9. What are the best practice examples of integrating technology into teacher education?

Strategies and methods (12%) are the second area of question area in terms of relevance. Examples of the main questions are the following:

1. How can prospective teachers apply the knowledge on alternative conceptions of heat and temperature in their teaching proposal?
2. In physics education, is a learning cycle a good strategy of teaching to improve the understanding of students?
3. What is an efficient approach to an astronomical observation for a lay audience?
4. What are the best learning approaches to teach physics in large student groups?
5. What strategies can I use to involve my students more during classes?
6. How do we encourage sense-making if we reward only the correct canonical answer?
7. Pursuant to what basic contents is it possible to articulate the science degree and the subsequent teaching so that, when actual teaching takes place, the teaching of science will be oriented towards an enquiry based on models?

Systems needed for teacher education and role of institutions generate 10% of questions, reproducing the spectra of research questions we found in the relevant EU inquiries.

Contents and contextualization in learning collect 10% of questions, as per the following:

1. Is it really necessary to cover the amount of content that is currently taught in secondary education?
2. How can we use cosmology context to teach advanced physics topics?
3. Is it helpful to start with concrete context like light bulb when teaching electricity?
4. To what extent can a teaching/learning sequence about spectra foster students' understanding of waves?
5. What contributions can the use of PSTU lend to the teaching of cosmology?
6. How can we build on chemistry knowledge to introduce basic quantum mechanics concepts?

Teaching specific contents is another area covered by questions:

1. What are the challenges in teaching Einsteinian physics at upper secondary school level?
2. What is the common order of teaching the concepts of force and momentum in your country? Do you agree with it, and why?
3. What is your experiences about teaching electromagnetic induction at secondary schools?
4. What makes the learning domain of general relativity challenging?
5. What should physics teachers know about entanglement?
6. Is it possible to introduce summative lectures that frame mechanics contents in a discipline-culture framework?

Laboratory role trigger off 7% of questions as follows:

1. Do you think laboratory practices foster significant learning?
2. In classes that involve concepts of modern and contemporary physics, do teachers use more experimental activities or just phenomenological descriptions?
3. What are student views regarding the implementation of online science laboratory work and what are the theoretical grounds of this method?
4. What is the role of laboratory activities to enhance students' interpretation of visual representations in physics?
5. Which experimental skills are required for physics ENEM items?

The other categories of Fig. 2 are self-explanatory. The following questions on research and education are particularly motivated:

1. To what degree should we require new teachers/university faculties to become familiar with the current state of education research before starting their professional practice?
2. What is the correct relationship between mentoring and research?
3. Which test was used for your research?

Very few are the questions posed for primary scientific education: only two, asking whether and how energy can be treated in primary education and asking about good practices in general.

Questions on university teacher education are few, but interesting, such as the following:

1. Is it recommended to have a research background to lecture physics at university level?
2. What common problems do engineering students encounter with physics assignments in the first year?
3. For a new university faculty, which presumably have extensive research experience, what specific training or preparation in pedagogy is offered to bring them up to speed in that regard?
4. Physics education also means preparing new researchers in physics. Most of researchers nowadays have to write codes to perform analyses. How much of that should be embedded in the physics curricula?

Questions announced as focused on pre-service teacher education cover a very widespread, as the following examples indicate:

1. How and how much statistics do students learn to become physics teachers?
2. How can we teach pre-service teachers to incorporate new learning strategies into practical assignments in classroom?
3. How can we teach pre-service teachers to use contextualized experiences in classroom?
4. How can we train students better in pre-service teacher education to initiate and moderate/lead discussions among scholars about problems in physics?
5. How do we encourage didactical reconstruction by trainees?

6. How do we train prospective teachers to inspire and motivate their students?
7. Which interdisciplinary science ideas do pre-service teachers need to know?
8. Which kind of aspects of theoretical physics/mathematical physics are relevant for those who become teachers compared to those who become physicists?
9. Which is the role of practical work in pre-service teacher education?
10. Which strategy does work best to improve PCK in physics regarding generalist pre-service primary teachers?
11. Do pre-service teachers have a strong sense of agency at the start of the course?

The 115 question posed and their analysis offer a scenario of the main needs of researchers and teachers in the primary, secondary and university levels and its organization herein described inspire the contributions on teacher education.

Acknowledgements I am very grateful to the organizers of the GIREP-MPTL Congress in San Sebastian, Jenaro Guisasola and Kristina Zuza for the idea of Dialogues and for the help. I thank GIREP for the learning opportunities offered. I thank the PLS-Fisica Project for the research opportunity granted in the teacher education area. I thank my research group for sharing the last 20 years of research in teacher education.

References

Abell S (2007) Research on science teachers' knowledge. In: Abell SK, Lederman NG (eds) Handbook of research on science education. Erlbaum, Mahwa, pp 1105–1149

Aiello ML, Sperandeo Mineo RM (2001) Educational reconstruction of physics content to be taught and of pre-service teacher training: a case study. IJSE 22(10):1085–1097

Amin TG, Smith C, Wiser M (2014) Student conceptions and conceptual change: three overlapping phases of research. In Lederman N, Abell S (eds) Handbook of research in science education, vol II. Routledge, p 4

Berger H, Eylon B-S, Bagno E (2008) Professional development of physics teachers. J Sci Educ Technol 17:399–409

Borko H (2004) Professional development and teacher learning. Educ. Res. 33(8):3–15

Bozzo G, Michelini M, Viola R (2010) Students and perspective teachers interpreting simple situation of induction phenomena. In Menabue L, Santoro G (eds) New trends in science and technology education. Selected papers, vol 1. CLUEB, Bologna, pp 355–363. ISBN 978-88-491-3392-9

Buchberger F, Campos BP, Kallos D, Stephenson J (2000) Green paper on teacher education in Europe. In: Thematic network on teacher education in Europe, Umea Universitaet. ISBN 91-973904-0-2

Corni F, Michelini M, Stefanel A (2004) Strategies in formative intervention modules for physics education of primary school teachers: a coordinated research in Reggio Emilia and Udine. In: Michelini M (ed) Quality development in the teacher education and training. Selected papers in Girep book. Forum, Udine, pp 382–386. ISBN: 88-8420-225-6

Davis A, Smithey J (2009) Beginning teachers moving toward effective elementary science teaching. Sci Educ 93(4):745–770

Dineke EH, Tigelaar et al (2004) The development and validation of a framework for teaching competencies in higher education. High Educ 48(2):253–268. JSTOR, https://www.jstor.org/stable/4151578. Accessed 1 July 2020

Duit R, Gropengießer H, Kattmann U (2005) Towards science education research that is relevant for improving practice: the model of educational reconstruction. In: Fischer HE (ed) Developing standards in research on science education. Taylor & Francis, London, pp 1–9

Elbaz F (1983) Teacher thinking: a study of practical knowledge. Nichols, New York, p 11

Eurydice (1998, 2003) The teaching profession in Europe: profile, trends and concerns. Key topics https://eacea.ec.europa.eu/nationalpolicies/eurydice/publications_en

Fedele B, Michelini M, Stefanel A (2005) 5–10 years old pupils explore magnetic phenomena in Cognitive Laboratory (CLOE). In: Pitntò R, Couso D (eds) CRESILS. Selected papers in ESERA Publication, Barcelona. ISBN: 689-1129-1

Leto F, Michelini M (2014) Energy transformations in primary school: outcomes from a research based experimentation. In: Tasar F (ed) Proceedings of the world conference on physics education 2012, Pegem Akademiel, pp 463–471. ISBN 978-605-364-658-7

Michelini M (2001) Supporting scientific knowledge by structures and curricula which integrate research into teaching. In: Pinto R, Surinach S (eds) Physics teacher education beyond 2000 (Phyteb2000), Girep book. Selected contributions of the Phyteb2000 international conference. Elsevier, Paris, p 77

Michelini M (2003) New approach in physics education for primary school teachers: experimenting innovative approach in Udine University. In: Gil EM (ed) Teaching physics for the future, A-37, SCdF. Selected papers of the VIII Inter-American conference on physics education, Havana, Cuba; Michelini M (2003) New approach in physics education for primary school teachers: experimenting innovative approach in Udine University. In: Ferdinande H, Vaicke E, Formesyn T (eds) Inquiries into European higher education in physics. European Physics Education Network (EUPEN), vol. 7, p l80. ISBN 90-804859-6-9

Michelini M (ed) (2004a) Quality development in the teacher education and training. In: Selected papers in Girep book. Forum, Udine. ISBN: 88-8420-225-6

Michelini M (2004b) Physics in context for elementary teacher training. In: Quality development in the teacher education and training

Michelini M (2004c) Physics in context for elementary teacher training. In: Michelini M (ed) Quality development in the teacher education and training. Selected papers in Girep book. Forum, Udine, pp 389–394. ISBN: 88-8420-225-6

Michelini M (2006) The learning challenge: a bridge between everyday experience and scientific knowledge. In: Planinsic G, Mohoric A (eds) Informal learning and public understanding of physics. Selected papers in Girep book, Ljubijana (SLO), pp 18–39. ISBN 961-6619-00-4

Michelini M (2010) Building bridges between common sense ideas and a physics description of phenomena to develop formal thinking. In: Menabue L, Santoro G (eds) New trends in science and technology education. Selected papers, vol. 1. CLUEB, Bologna, pp 257–274. ISBN 978-88-491-3392-9

Michelini M, Mossenta A (2014) Building a PCK proposal for primary teacher education in electrostatics, in teaching and learning physics today: challenges? Benefits? In: Kaminski W, Michelini M (eds) Selected paper books of the international conference GIREP-ICPE-MPTL 2010, Reims 22–27 Aug 2010, Lithostampa, Udine, pp 164–173. 978-88-97311-32-4

Michelini M, Sperandeo Mineo RM (2014) Challenges in primary and secondary science teachers education and training. Kaminski W, Michelini M (eds) Teaching and Learning Physics today: challenges? Benefits? Selected paper books of the international conference GIREP-ICPE-MPTL 2010, Reims 22–27 Aug 2010, Udine: Lithostampa, pp 143–148. 978-88-97311-32-4

Michelini M, Stefanel A (2008) Secondary teachers discussing the pedagogical and cultural aspects in teaching/learning quantum physics. In: Sidharth BG (ed) Frontiers of fundamental and computational physics—FFP9

Michelini M, Stefanel A (2015) Research based activities in teacher professional development on optics. In: Fazio C, Sperandeo RM (eds) Proceedings of Girep Conference, Palermo 2015, Palermo, pp 853–862

Michelini M., Stefanel A (2016) Conceptual lab of operative exploration (Cloe) as research contexts to explore pupils reasoning in physics. In: IOP proceedings of Kralow GIREP seminar

Michelini M, Vidic E (2016) Research based experiment on the concept of time for scientific educa-tion on transversal perspective in primary school, communications to the HSCI 2016 congress, Brno 18–22 July 2016. In Martin Costa MFPC, Dorrio JBV, Trna J, Trnova E (eds) Hans-on: the heart of the science education, p 164

Michelini M, Rossi PG, Stefanel A (2004) The contribution of research in the initial teacher forma-tion. In: Michelini M (ed) Quality development in the teacher education and training. Selected papers in Girep book. Forum, Udine, pp 166–172. ISBN: 88-8420-225-6

Michelini M, Santi L, Stefanel A, Vercellati S (2011) Community of prospective primary teachers facing the relative motion and PCK analysis. In: Teaching and learning physics today: challenges? Benefits? International conference GIREP-ICPE-MPTL 2010 proceedings, Université de Reims Champagne Ardenne, Reims 22–27 Aug 2010. http://www.univ-reims.fr/site/evenement/girep-icpe-mptl-2010-reims-international-conference/gallery_files/site/1/90/4401/22908/29321/29840.pdf

Michelini M, Santi L, Stefanel A (2013) La formación docente: un reto para la investigación, Revista Eureka sobre Enseñanza y Divulgación de las Ciencias 10 (Núm. Extraordinario), pp 846–870

Michelini M, Santi L, Stefanel A (2014) PCK approach for prospective primary teachers on energy. In: Tasar F (ed) Proceedings of the world conference on physics education 2012, Pegem Akademiel, pp 473–477. ISBN 978-605-364-658-7

Park S, Oliver SJ (2008) Revisiting the conceptualization of pedagogical content knowledge (PCK): PCK as a conceptual tool to understand teachers as professionals. Res Sci Ed 38:261–284

Pinto R, Surinach S (eds) (2001) Physics Teacher Education Beyond 2000 (Phyteb2000). In: Girep book of Selected contributions of the Phyteb2000 international conference in Barcelona. Elsevier

Sassi E, Michelini M (2014a) Physics Teachers' Education (PTE): problems and challenges, in frontiers of fundamental physics and physics education research. In: Burra GS, Michelini M, Santi L (eds) Book of selected papers presented in the international symposium frontiers of fundamental physics, Udine 21–23 Nov 2011, 12th edn. Springer, Cham, pp 41–54. 978-3-319-00296-5

Sassi E, Michelini M (2014b) Physics teachers' education (PTE): problems and challenges, in frontiers of fundamental physics and physics education research. In: Burra GS, Michelini M, Santi L (eds) Book of selected papers, Springer, Cham, Heidelberg, NY, Dordrecht, London [978-3-319-00296-5], pp 41–54

Shulman LS (1987) Knowledge and teaching of the new reform. Harvard Educ Rev 57:1–22

Sperandeo Mineo RM, Fazio C, Tarantino G (2006) Pedagogical content knowledge development and pre-service physics teacher education: a case study. Res Sci Educ 36:235–268

Tasar F (ed) Proceedings of the world conference on physics education 2012, Pegem Akademiel. ISBN 978-605-364-658-7

Testa I, Michelini M (2007) Prospective primary teachers 'functional models of electric and logic circuits: results and implications for the research in teacher education, in Van den Berg E (ed) Modelling in physics and physics education

Vercellati S, Michelini M (2014) Electromagnetic phenomena and prospective primary teachers. In: Dvořák L, Koudelková V (eds) Active learning—in a changing world of new technologies. Charles University in Prague, Matfyzpress Publisher, Prague, pp 235–241. 978-80-7378-266-5

Vidic E, Michelini M, Maurizio R (2020) Research based intervention module on fluids for prospective primary teachers. In: San Sebastian conference GIREP-MPTL Publications, in press

Vosniadou S (2008) International handbook of research on conceptual change. Routledge, New York. ISBN 978-0-8058-6044-3

Primary, Secondary and University Pre-service Physics Teacher Education—What Scientific Education Is Relevant for Becoming a Physics Teacher in a Technological World?

Knut Neumann

Abstract There is no doubt that teachers play a central role sparking students' interest in physics and supporting them in learning about physics. The question is, however, what teachers need to know and, more importantly, what other qualities teachers need to possess in order to meet the demands that come with that central role. In this position paper, I will address questions about the main issues we are currently facing in physics teacher education and what can be done about it. I will argue for a stronger focus on non-cognitive qualities of teacher professional competence, their role in the organization of high-quality physics instruction and how these qualities can be developed in physics teacher education.

1 Introduction

Teachers play a central role in organizing high-quality instruction and thus student learning (Hattie 2009). But what do teachers need to know or, more generally, be capable of in order to organize high-quality instruction? In his attempt to answer this question, Lee Shulman introduced the idea of teacher professional knowledge, including content knowledge, as well as pedagogical knowledge, but most importantly pedagogical content knowledge (Shulman 1987). Shulman (1987) characterized pedagogical content knowledge as the amalgam of content and pedagogy (NOT: content knowledge and pedagogical knowledge); that is, the special knowledge a teacher needs to teach content (the subject matter) to students. Pedagogical content knowledge should therefore not be mistaken for something that emerges from bringing high content and pedagogical knowledge to teaching but something that requires explicit instruction. In short: Teacher education needs to support (future) teachers in developing content knowledge (CK), pedagogical knowledge (PK) and pedagogical content knowledge (PCK).

K. Neumann (✉)
Leibniz-Institute for Science and Mathematics Education (IPN), Kiel, Germany
e-mail: neumann@leibniz-ipn.de

© The Editor(s) (if applicable) and The Author(s), under exclusive license to Springer Nature Switzerland AG 2020
J. Guisasola and K. Zuza (eds.), *Research and Innovation in Physics Education: Two Sides of the Same Coin*, Challenges in Physics Education,
https://doi.org/10.1007/978-3-030-51182-1_4

53

Much research has shown, however, that successful interaction in a complex environment requires more than just knowledge (White 1959). Teacher beliefs may, for example, substantially influence teachers' performance in class (Bryan 2012), and a recent study suggests that whereas teacher professional knowledge (i.e. PCK) may have a strong effect on student learning, teacher motivation (i.e. the interest in teaching the subject) has the stronger effect on the development of students interest (Keller et al. 2017). Baumert et al. (2013) therefore suggested a model of teacher professional competence that includes three more qualities of teachers in addition to teacher professional knowledge, namely teacher beliefs and values, teacher motivational orientation and teacher self-regulatory skills (Fig. 1). This model aligns well with the more recently published (Revised) Consensus Model of Teacher Professional Competence (for details see Neumann et al. 2018). However, the latter defines more explicitly the role of the non-cognitive aspects of teacher professional competence in terms of amplifiers and filters that affect the development of teacher professional knowledge and how it plays out in teaching. In terms of the development of teacher professional competence, the Model of Teacher Professional Competence by Baumert and colleagues may be interpreted to (implicitly) incorporate a developmental perspective in that teachers in Germany are usually to develop CK, PCK and PK at university separately in physics, physics education and pedagogy classes.

Beliefs and values, motivational orientations or self-regulatory skills are typically not (explicitly) targeted. Only when teachers reach the in-service training or induction phase both cognitive and non-cognitive aspects are attended too. However, often more informally, as a part of the mentoring by more experienced teachers. This results in the first two questions about teacher education that I want to discuss in this paper: (1) What do we know about the development of prospective teachers'

Fig. 1 Model of teacher professional competence (Baumert et al. 2013)

professional competence and (2) what are learning opportunities for prospective teachers to develop professional competence?

Obviously, physics and physics education classes are major learning opportunities for prospective teachers to develop CK and PCK during university teacher education (Neumann et al. 2017). A central part of physics education at the university is physics lab work classes in which physics students are expected to develop the required inquiry skills, or, more specifically, experimentation skills required to become an (experimental) physicist. But why should prospective teachers be prepared to become (experimental) physicists? It appears reasonable to assume that school physics instruction requires a different skill set than experimental physics. This leads to the next question: (3) How can prospective physics teachers be prepared to do lab work in schools? There has been much research on inquiry learning focusing on both the role of the teacher in organizing inquiry learning and students' inquiry learning itself. This research has the potential to inform the design and organization of lab work classes for prospective teachers. On a more general level, the other question I want to discuss in this context is: (4) How can physics education research be integrated in the education of prospective physics teachers?

In order to improve the education of (prospective) physics teachers, we need to be able to closely monitor the development of teacher professional competence from university to the in-service training/induction phase. The fact that we are focusing on different aspects of prospective teachers' professional competence at different stages of teacher education leads to the last question I want to discuss: (5) How can we evaluate physics teaching competence? I will discuss these questions in the following three sections drawing on examples from my own research on prospective physics teachers' education. I will conclude this paper by addressing one more question in the discussion, which is: What are the most pressing issues in research on the education of (prospective) physics teachers that need to/should be addressed in the years to come.

2 The Development of Prospective Teachers' Professional Competence

The trinity of content knowledge (CK), pedagogical content knowledge (PCK) and pedagogical knowledge (PK) has developed into the main framework used to describe the first stage in the development of prospective teachers' professional competence. During university teacher education programs, prospective physics teachers are commonly expected to develop CK as a result of taking physics classes, PCK as a result of taking physics education classes and PK as a result of pedagogy classes. For a long time, the question in research on the development of teacher professional knowledge has been to which extent or how the development of these knowledge bases affect or depend on each other, respectively. One major finding was that CK has been found to be a prerequisite for the development of PCK in mathematics (Baumert et al.

2013). Recent research on (prospective) physics teachers' professional knowledge has confirmed a tight correlation between CK and PCK (Kirscher et al. 2016; Riese and Reinhold 2012). In a more differentiated study of prospective physics teachers at different stages in their university teacher education program, (Sorge et al. 2017) have found that the correlation between CK and PCK is lower for beginning teachers, $r = 0.60, p < 0.01$, than the correlation between PCK and PK, $r = 0.94, p < 0.01$. For advanced pre-service teachers, the correlation between CK and PCK was substantially higher, $r = 0.89, p < 0.001$, than between PCK and PK, $r < 0.69, p < 0.001$. This suggests that (the structure of) prospective physics teachers changes throughout university teacher education. It seems that in the beginning, prospective physics teachers' professional knowledge is mainly general pedagogical knowledge. The development of advanced prospective teachers' professional knowledge, in contrast, seems to be mainly driven by their CK, as CK and PCK are substantially more correlated for advanced than for beginning prospective physics teachers. Interestingly, research shows that this observation applies mainly to secondary level teachers, whereas primary level teachers' professional knowledge mainly comprises PCK and PK (Möller et al. 2013). Interestingly, little is known about university level physics teachers' professional knowledge. While it may be safe to assume that university level teachers possess a high content knowledge, since most of them only had a pure academic education (with no classes in physics education or pedagogy), it is unclear if they possess any PCK or PK beyond intuitive knowledge and if the knowledge they have suffices to ensure high-quality teaching. In terms of learning opportunities, the mentioned findings support the assumption that CK is mainly developed in physics classes, PCK in physics education classes and PK in pedagogy classes as students usually take physics and pedagogy classes from the very beginning of their university teacher education program, but in many cases, physics education classes are supposed to be taken only later in the program. Another interesting finding by Sorge et al. (2017) was that the development of CK and PCK was also driven by observation of school instruction during practical phases supporting the common assumption that practical phases can foster the development of professional knowledge.

3   Rethinking the Education of Prospective Physics Teachers

In terms of the development of more specific abilities or skills such as the skills required to organize (high-quality) practical work, we need to focus on specific classes. Prospective teachers (should) develop the skills to organize high-quality practical work in physics lab work classes. At least in Germany, prospective physics teachers often attend the same classes as physicists. These classes are, in my opinion, subpar learning opportunities for physics teachers at best since they ideally focus on providing students with the skills needed to become experimental physicists. These skills are, however, substantially different from the skills needed by physics teachers. Students are, for example, sometimes learning how to use the software LABVIEW to analyse their data. But LABVIEW is not or at least not commonly utilized in schools. In fact, many schools use software specifically designed for students (e.g. CASSY). As a consequence of the mismatch in the goals pursued in lab work classes for physics students and the goals pursued by prospective physics teachers, the students perceive the lab work class as not or not sufficiently relevant for their future work as teachers. The result likely is a lack of motivation that in turn likely leads to limited learning. In order to address these issues and to improve prospective physics teachers' development of professional knowledge, at the University of Kiel physics, lab work classes were designed and implemented specifically for teachers (Andersen et al. 2018). The underlying assumption of doing so was that a lab work class with experiments specifically designed for prospective teachers would raise the perceived relevance and thus their motivation, which in turn would positively affect their learning (i.e. knowledge gain). Interestingly, we found some variance in how relevant prospective teachers perceived the different experiments and respective differences in their learning (Fig. 2). While this confirms that providing prospective teachers with learning opportunities specifically designed from them can increase their learning, more research is needed about which features determine the relevance perceived.

How this research can be incorporated into the education of physics teacher is a major question for physics education researchers and physics educators alike. The (revised) consensus model of teacher professional competence delineates three types of PCK that prospective teachers are expected to acquire throughout their career: collective PCK (or cPCK), personal PCK (or pPCK) and enacted PCK (or ePCK). Sorge, Stender and Neumann (2019) argue that cPCK is commonly developed throughout university teacher education, whereas pPCK and ePCK are developed

Fig. 2 Relationship between perceived relevance of experiments in physics lab work course for prospective teachers and respective knowledge gain (Andersen et al. 2018)

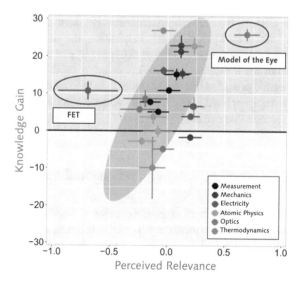

Fig. 3 Sample seminar
incorporating PCK (cPCK)
as produced by recent
physics education research

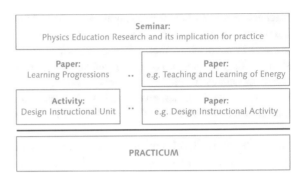

through teaching—in practical phases and, more importantly, everyday life teaching. Since cPCK is characterized as the knowledge commonly shared by teachers and researchers, cPCK is the knowledge generated through research. As a result, this knowledge needs to be incorporated into university teacher education.

But how can university teacher education classes incorporate recent knowledge developed through research without a need to consistently redesign classes? At Kiel University, in order to incorporate findings from most recent physics education research, we run a seminar in which students are introduced to physics education research, typical physics education research paradigms, methodological approaches and analysis methods at the example of recent research papers (Fig. 3). After an introduction, in each class session, a group of students presents the core findings from a recent research. In addition, each group of students presents an activity that implements the findings presented in the paper. This way students need to (a) think more deeply about the impact of the research they have read about for physics instruction and (b) provide their fellow students with the opportunity to explore the implications first hand. The seminar is supposed to prepare students for an eight-week practicum at schools, they are taking immediately afterwards in order to foster the compilation of cPCK into pPCK/ePCK. In addition, the seminar provides students with fundamental knowledge about and skills in physics education research and thus prepares them for a research career (i.e. doing a Ph.D.). A preparation that is desperately needed provided that many students in teacher education programs envision a career as a teacher but do not even know about a potential career as an academic (i.e. physics education researcher).

4   Assessing Teacher Professional Competence

Teacher professional competence, as discussed above, develops in several stages. In order to be able to monitor prospective teachers and teacher competence development for formative and summative purposes, instruments are needed that can reliably and validly assess teacher professional competence (or aspects thereof). The main focus of university teacher education is prospective teachers' professional knowledge; that is their CK, PCK and PK. Instruments that have successfully been used for assessing teacher professional competence at this stage (i.e. teacher professional knowledge) are paper and pencil tests (Sorge et al. 2017), content representations (Hume and Berry 2011) or lesson plans (de Jong and van Driel 2004). Paper and pencil tests have been used to assess specifically collective PCK, whereas content representations and lesson plans have been used to assess more personal aspects of prospective teachers PCK (i.e. pPCK and ePCK).

Instruments that have successfully been used in the induction phase to assess the compilation of cPCK into pPCK/ePCK, and hence, the development of teacher professional knowledge into professional competence includes lesson plans (Sorge et al. 2019; de Jong and van Driel 2004) and content representations (Hume and Berry 2011). In addition, pedagogical and professional-experience Repertoires (PaPeRs) have been used to assess ePCK as a representation of the amalgam of other knowledge bases (and hence professional competence) (Betram and Loughran 2012). In order to assess teachers' professional competence of in-service, experienced teachers video vignettes (Meschede et al. 2017) and video analysis procedures (Fischer et al. 2014) have been utilized in order to capture this highly refined non-accessible ePCK or, actually, teacher professional competence.

Overall, there is a wide range of instruments available to assess different aspects of teacher professional competence or teacher professional competence as a whole. The main point is to choose the instrument based on which aspect or level of compilation is supposed to be assessed, in order to achieve not only reliability but maximum validity of the findings.

5   Discussion

In summary, recent research has provided substantial insights into prospective teachers' CK, PCK and PK, the relationship between these knowledge bases and how they develop. Substantially, less is known about the specific learning opportunities that lead to the development of these knowledge bases or how specific learning opportunities designed to improve physics teacher education. Even less is known about how teacher professional knowledge develops into teacher professional competence; or, more specifically, the role of non-cognitive aspects in this process. The (revised) consensus model of teacher professional competence describes the development of teacher professional competence as the compilation of cPCK into pPCK/ePCK (Sorge et al. 2019). The model also notes that this process is influenced by non-cognitive aspects which act as amplifiers or filters, respectively. However, so far little research has been presented that sheds light on the specific role of non-cognitive aspects. Research that does focus on the role of non-cognitive aspects rather shows that these aspects may have a considerable influence on instruction in addition to teacher professional knowledge. Keller, Neumann and Fischer (2017), for example, found that while teacher PCK affects students learning (i.e. gains in student achievement in a content knowledge test), teacher motivation was a main predictor

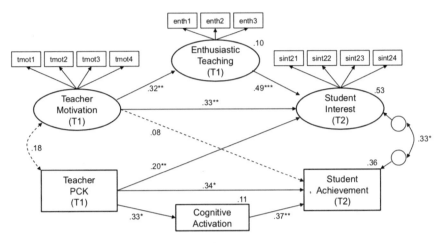

Fig. 4 Role of cognitive and non-cognitive aspects of teacher professional competence for the development of student achievement and interest (Keller et al. 2017)

of the development of student interest. Both effects were mediated by instruction confirming that cognitive and non-cognitive aspects of teacher professional competence need to be considered if we aim to better understand what (prospective) physics teachers need in order to create high-quality instruction (Fig. 4). In summary, I think that while we need more research understanding how prospective teachers' professional knowledge can best be developed throughout teacher education and how its compilation into teacher professional competence can be supported, we also need to further explore the role of non-cognitive aspects in this process but also in their relevance for organizing high-quality instruction.

References

Andersen J, Block D, Neumann K and Wendlandt H 2018 Lehramtsausbildung Physik 2.0: Vernet-zung von Fach, Fachdidaktik und schulpraktischen Aspekten. In: Brouër B, Burda-Zoyke A, Kilian J, Petersen I (eds) Vernetzung in der Lehrerinnen- und Lehrerbildung: Ansätze, Methoden und erste Befunde aus dem LEAP-Projekt an der Christian-Albrechts-Universität zu Kiel. Waxmann, Münster, pp 23–36

Baumert J, Kunter M, Blum W, Klusmann U, Krauss S, Neubrand M (2013) Cognitive activation in the mathematics classroom and professional competence of teachers. Springer, New York

Betram A, Loughran J (2012) Res Sci Educ 42(6):1027–1047

Bryan LA (2012) Research on science teacher beliefs. In: Fraser BJ, Tobin KG, McRobbie CJ (eds) Second international handbook of science education, Chapter 33. Springer, New York, pp 477–95

de Jong O, van Driel J (2004) IJSME 2(4):477–491

Fischer HE, Neumann K, Labudde P, Viiri J (2014) Quality of instruction in physics: comparing Finland Switzerland and Germany. Waxmann, Münster

Hattie J (2009) Visible learning. Routledge, London

Hume A, Berry A (2011) Res Sci Educ 41:341–355

Keller M, Neumann K, Fischer HE (2017) J Res Sci Teach 54:586–614

Kirscher S, Borowski A, Fischer HE, Gess-Newsome J, von Aufschnaiter C (2016) Int J Sci Educ 38:1343–1372

Meschede N, Fiebranz A, Möller K, Steffensky M (2017) Teach Teach Educ 66:158–170

Möller K, Kleickmann T, Lange K (2013) Naturwissenschaftliches Lernen im Übergang von der Grundschule zur Sekundarstufe. In: Fischer HE, Sumfleth E (eds) nwu-essen: 10 Jahre Essener Forschung zum naturwissenschaftlichen Unterricht, Chapter 2. Logos, Berlin, pp 57–112

Neumann K, Härtig H, Harms U, Parchmann I (2017) Science teacher preparation in Germany. In: Pedersen J, Isozaki T, Hirano T (eds) Model science teacher preparation programs. Information Age Publishing, Greenwich, CT, pp 29–52

Neumann K, Harms U, Kind V (2018) Int J Sci Educ. https://doi.org/10.1080/09500693.2018.149 7217

Riese J, Reinhold P (2012) Z Erziehwiss 15:111–143

Shulman LS (1987) Harvard Educ Rev 57:1–22

Sorge S, Kröger J, Petersen S, Neumann K (2017) Int J Sci Educ. https://doi.org/10.1080/09500693. 2017.1346326

Sorge S, Stender A, Neumann K (2019) The development of teachers' professional competence. In: Hume A, Cooper R, Borowski A (eds) Repositioning pedagogical content knowledge in teachers' professional knowledge. Springer, New York, pp 149–164

White RW (1959) Psychol Rev 66:297–333

Concerns About Relevant Physics Education in a Technological World: An Overview of GIREP Participants' Questions

María Gabriela Lorenzo⊙

After some 30 years of [analysing teaching], I have concluded that classroom teaching … is perhaps the most complex, most challenging, and most demanding, subtle, nuanced and frightening activity that our species ever invented.
Lee Shulman (2004, p. 504)

Abstract Physics teachers have a variety of concerns about how to teach physics in a technological world. In this paper, an analysis of these ideas is presented and discussed. In the framework of a meeting, participants were invited to formulate some questions to be debated by a panel of experts during the event. A content analysis was performed using both qualitative and quantitative simple approaches in order to detect teachers' concerns with the aim to promote a subsequent reflection. Most of the questions referred to how to deal with teachers' education and their professional knowledge development. The analysis of the question applying PCK model showed a strong interest of the physics teachers on topic-specific knowledge and the best strategies to teach it. This meeting was a formidable opportunity to detect different needs and requirements that will be a great contribution in order to rethink physics and science education in this technological world.

1   Introduction

The purpose of this work is to discuss the main concerns of physics teachers from a critical analysis of a set of questions posed by a group of physics teachers who participated in an international congress on physics education.

M. G. Lorenzo (✉)
Universidad de Buenos Aires, Facultad de Farmacia y Bioquímica, Consejo Nacional de Investigaciones Científicas y Técnicas, Junín 956, CP1113 Ciudad Autónoma de Buenos Aires, Argentina
e-mail: glorenzo@ffyb.uba.ar

J. Guisasola and K. Zuza (eds.), *Research and Innovation in Physics Education: Two Sides of the Same Coin*, Challenges in Physics Education,
https://doi.org/10.1007/978-3-030-51182-1_5

Research and innovation in physics education: two sides of the same coin Conference was organized by Groupe International de Recherche sur l'Enseignement de la Physique (GIREP) and Multimedia in Physics Teaching and Learning (MPTL). It took place in Donostia-San Sebastián City, in July 2018. Its foremost aim was to offer the attendees the opportunity to share their ideas and experiences, as well as to get in contact with the advances in the realm of physics education. The conference was organized in seven strands:

#	Strands	Code
1	Physics teaching and learning at primary and secondary education	P&S
2	Physics teaching and learning at university	U
3	Pre-service and in-service physics teachers education	TE
4	Physics education in non-formal setting	NFS
5	Physics into STEM teaching and learning	STEM
6	ICT and multimedia in physics education	ICT
7	Nature of science, gender and sociocultural issues in physics education	NOS

The concepts and reflections exposed here are the result of the study about the interaction between the attendees and a panel of experts, through a particular activity named *dialogues*. This work aims to give a feedback on the proposed questions to *Dialogue 1. Primary, secondary and university pre-service physics teacher education. What scientific education is relevant for becoming physics teacher in a technological world?*

To detect the main concerns of the questions suggested by the participants, a content analysis was performed using both qualitative and quantitative simple approaches.

The leader of the dialogue 1, Marisa Michelini, posed a set of questions in order to organize the discussion during the face-to-face interchange (Fig. 1), and these were taken into account to do the analysis.

2 Contributions to the Debate

2.1 The Questions and the Strands

The inquiry was based on these initial interrogations: What did the collected questions show about physics teachers' concerns? Can some "requirements" be recognized and distinguished from "needs"? How could this contribute to improve physics teachers' education?

In a first step, 113 questions could be categorized as part of one of the proposed strands (Fig. 2). It shows that the 40.71% were related with teachers' education and their professional knowledge development.

1) We have three different teacher profiles: primary teachers, secondary school teachers and university teachers. From your research evidence,
   a) Which are the main problems for each profile? How you find solutions?
   b) How sort of activities in teacher education promote their professional competencies? How?
   c) How can be integrated the different competencies needed by teachers in the different levels considered?
2) How prospective teachers can find opportunities to develop the main professional skills
   a) Which is the role of psychological -pedagogical education?
   b) How can be discussed the contents for teaching conceptual competences?
   c) Which kind of lab emerge to be useful?
   d) How to conduct apprenticeship?
3) How can be prepared Physics teacher for a significant integration of:
   a) ICT in school activities?
   b) Lab work in school activities?
4) How to integrate Physics education research in Physics teacher education?

Fig. 1 General question of dialogue 1

Fig. 2 Percentage of questions categorized by strands

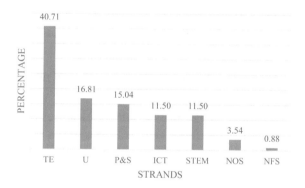

It is important to note that the second group of questions were about teaching and learning physics at the university level in undergraduate courses (Guisasola 2019). This is an indicator of the relevance of this group of students as a new branch of science education.

The technological issues were mentioned sometimes as integrated with other contents in a STEM point of view or as a digital resource to improve or modify teaching or learning. The lack of questions about the nature of science and the involved metadisciplines in science education as well as gender and sociocultural issues have disclosed a relevant point to be considered by researchers, trainers and tutors in order to enhance and boost physics teachers' education. On the same direction, to think about plausible interactions between non-formal settings and teachers' education is another issue to pay attention to.

Fig. 3 Different questions about teacher education

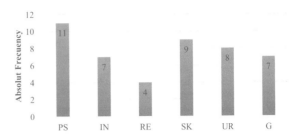

Going into depth in the first strand, the questions could be classified into six new types: pre-service physics teachers' education (PS), in-service (IN), resources (RE), skills (SK), university, research (UR), general aspects (G) (Fig. 3).

Some selected question could display the interest of the participants:

About pre-service and in-service teacher education there were concerns about different strategies, contextualized experiences, self-confidence, some particular complex contents, among others:

> How can we [make] pre-service teachers learn to incorporate new learning strategies into practical assignments in the classroom? (PS)
>
> What is the role of [the] practical work in pre-service teacher education? (PS)
>
> How can we support teachers to teach concepts of quantum physics? (IN)

Other questions pointed to some particular resources or some kind of teachers' abilities:

> The biggest question for young teachers is: how to teach without technology? (RE)
>
> Do we really want to develop critical thinking in teachers, and how to proceed in this regard? (SK)

Completing the picture, there were several reflective questions that pointed out the difference between teachers and physicist education, and the specific education of those who will be responsible to teach teachers.

> What aspects of theoretical physics/mathematical physics are relevant for those who become a teacher compared to those who become a physicist? (UR)
>
> What kind of training/knowledge (if any) should a physics teachers' educator have? (G)

2.2 About Topics of Interest

An analysis of the questions looking for particular topics of interest could show several specific subject-matter contents, the interaction with other disciplines and in a minor degree, the connection with other areas of knowledge (Fig. 4).

Specific Subject-matter contents	Other disciplines	Broader realm of knowledge
Quantum mechanics	Astronomy	Culture
Heat and temperature	Cosmology	Multidisciplinary world
Energy	Maths	Interdisciplinary ideas
Electricity	Statistics	History of Science
Spectroscopy	Chemistry	Epistemology
Force	Engineering	
Electromagnetism	Technology	
	Robotics	
	Computational issues	
	Others sciences	

Fig. 4 Topics of interest

Criteria	Questions
• Orientation to teaching Science	10
• Knowledge of Science Curricula	29
• Knowledge of Students' Understanding of Science	5
• Knowledge of Assessment of Scientific Literacy	4
• Knowledge of Instructional Strategies	18
• Relation School/University	3
• Physics Education Research	9

Fig. 5 Question analysis based on PCK model

2.3 Examining PCK Beneath the Questions

The Pedagogical Content Knowledge Model (PCK) (Shulman and Shulman 2004) offers an interesting and alternative approach to analyse the ideas of the participants of GIREP. This particular kind of knowledge is responsible for the professional teachers' development. Briefly, it could be considered as a mixture of other types of knowledge such as subject-matter knowledge, pedagogical knowledge, knowledge of student learning (and presumably others). So, it could help to inquire about which PCK components were included in the questions.

Using Magnusson et al. proposal (Magnusson et al. 1999) (Fig. 5), a new perspective was reached.

This table shows that specific subject matter of science curricula (as it has been presented in Fig. 4) and the searching of instructional strategies are the two most exposed concerns which could be found in the questions. This might be interpreted as a traditional vision of teaching centred in contents and in the application of proved recipes or "the best practices examples". In addition, the short number of questions about psychological processes involved in students' understanding (learning, motivation, metacognition) and in general scientific literacy (presumably related with metadisciplinar knowledge) go on the same direction.

3   Final Consideration and Perspectives

Teachers' education practices must be reviewed if a real improvement wants to be reached. For this purpose, there exist three important highlights to be considered:

- The articulation between school and university in order to reduce the gap between them and to promote a fruitful interchange. Consequently, possible new articulations would be generated among teachers, scholars and researchers in order to constitute mixed teams.
- In a similar sense, the learning community model is a powerful opportunity for teachers' trainings (Lantz-Andersson et al. 2018) and for the long professional learning along life.
- Science education research (Nussbaum 2017) must be an engine that impulses the previous points in order to investigate real problems to give appropriate solutions. The scientific literature will contribute to find a common language and a common interest in multilevel research teams where teachers, scientists and researchers could interact among themselves. In this way, the teaching task will be transformed into an object of investigation for teachers.

At last, a divorce between subject-matter knowledge and pedagogical knowledge exists in most of the formative courses. Therefore, to face the generalized idea that the teacher is who must blend both types of knowledge, it is necessary to convert teacher education in a blending proposal that enact *to teach teaching* (Lorenzo 2012). So, it is a must to give more participation to specific science educators allowing them to teach disciplinary courses within formal pre- and in-service teachers education.

This meeting was a formidable opportunity to detect different needs and requirements that will be a great contribution in order to rethink physics and science education in this technological world.

Acknowledgements Science and Technology Projects: PIP 11220130100609CO (2014–2016); FONCYT-PICT-2015-0044 & UBACYT-2018-20020170100448BA (2014–2017).

References

Guisasola J (2019) Research-based alternatives to traditional physics teaching at university and college. In: Pietrocola M (ed) Upgrading physics education to meet the needs of society. Springer, Cham, pp 127–132

Lantz-Andersson A, Lundin M, Selwyn N (2018) Twenty years of online teacher communities: a systematic review of formally-organized and informally-developed professional learning groups. Teach Teach Educ 75:302–315

Lorenzo MG (2012) Los formadores de profesores: El desafío de enseñar enseñando [Teachers' trainers: the challenge of to teach teaching]. Prof Rev Curriculum Form Prof 16(2):343–360. https://www.ugr.es/~recfpro/rev162COL3.pdf

Magnusson S, Krajcik J, Borko H (1999) Nature, sources, and development of pedagogical content knowledge for science teaching. In: Gess-Newsome J, Lederman NG (eds) Examining pedagogical content knowledge. Springer, Dordrech, pp 95–132

Nussbaum L (2017) Doing research with teachers. In: Moore E, Dooly M (eds) Qualitative approaches to research on plurilingual education. Research-publishing.net, Dublin, pp 46–67)

Shulman L (2004) Professional development: leaning from experience. In: Wilson S (ed) The wisdom of practice: essays on teaching, learning, and learning to teach. Jossey-Bass, San Francisco, pp 503–522

Shulman L, Shulman J (2004) How and what teachers learn: a shifting perspective. J Curriculum Stud 36:257–271

Critiquing Explanations in Physics: Obstacles and Pedagogical Decisions Among Beginning Teachers

Laurence Viennot and Nicolas Décamp

Abstract This chapter summarises a series of studies on the critical responses of beginning teachers when presented with questionable explanations in physics. Based on a typology of these responses and some hypotheses on the main factors of critical passivity, suggestions are made to facilitate the activation of a critical attitude in this population. The document recommends enlightening the pedagogical choices of explanations through a multi-criteria analysis. The final discussion draws attention to the crucial importance of the simplicity of an explanation as a criterion of choice for beginning teachers, even at the expense of its coherence.

1 Introduction

Given the widely accepted need to develop critical thinking in physics students and teachers, we conducted a series of five studies of beginning teachers' critical responses to explanatory physics texts (see, Viennot and Décamp 2018a; Viennot and Décamp 2018b for overview). The objective of these studies was to document links between critical attitude and conceptual development in the context of a number of physics topics (Décamp and Viennot 2015; Viennot and Décamp 2016a; Viennot and Décamp 2016b; Viennot and Décamp 2018c). The studies targeted texts commonly used in physics education or to explicate phenomena in academic settings and/or popularised accounts. This brief summary of the main findings highlights obstacles that may block the activation of a critical attitude in this population and considers the implications for physics teachers' practice. The chapter concludes with a discussion of objectives for teacher preparation and future research.

L. Viennot (✉)
Université de Paris, MSC, UMR 7057, Paris 75013, France
e-mail: laurence.viennot@gmail.com

N. Décamp
Université de Paris, LDAR, Universités d'Artois, Cergy-Pontoise, Paris-Est Créteil, Rouen, Paris 75013, France

J. Guisasola and K. Zuza (eds.), *Research and Innovation in Physics Education: Two Sides of the Same Coin*, Challenges in Physics Education,
https://doi.org/10.1007/978-3-030-51182-1_6

2 Critical Attitude: Typology and Obstacles

2.1 A Typology of Critical Attitude: Expert Anaesthesia, Delayed Critique, Early Critique

In discussing possible links between critical attitude and conceptual expertise, it might be expected that conceptual mastery in a given domain would entail an ability to criticise contestable explanations in that domain. However, our observations suggest that some individuals remain critically passive when asked to detect the flaws in inconsistent or logically incomplete explanations of physics phenomena despite possessing the means to do so. In one telling example, a hot air balloon was characterised as an isobaric situation. Although this contradicts a fundamental principle of fluid statics—that is, the role of pressure gradients in flotation—most textbook writers and physics teachers seemed to accept this hypothesis as valid (Viennot 2006).

As another example, radiocarbon dating is often incompletely explained; despite the decay of radiocarbon, the composition of the atmosphere is considered constant in time. In our investigation of this topic (Décamp and Viennot 2015), some interviewees with mastery of this topic failed to react to this logically incomplete explanation. It is also the case that popular accounts consistently fail to explain this point as if it were not an essential node of the explanation. We characterise cases of this kind, where conceptual mastery and critical passivity co-occur, as 'expert anaesthesia'.

In other cases, referred to as 'delayed critique', critical passivity was associated with defective conceptual mastery of the domain in question. Here, someone who is not an expert expresses a need to know more before offering any critique, even though no specialised knowledge is required. For instance, without any expert knowledge of capillarity, one might question the use of the diagram in Fig. 1 to demonstrate Young's relationship ($\gamma_{LG} \cos \theta = \gamma_{SG} - \gamma_{SL}$) for an angle of contact θ between solid (S) and liquid (L) in the presence of a gas (G) and the interfacial tension coefficients γ_{LG}, γ_{SG} and γ_{SL}. This diagram is frequently used as a free body diagram, in which forces (by unit length) are seen to act on an immaterial line. This is meaningless in Newtonian theory, yet beginning teachers were often very slow to articulate this point when

Fig. 1 Diagram introducing Young's formula ($\gamma_{LG} \cos \theta = \gamma_{SG} - \gamma_{SL}$), which could be criticised without expert knowledge of capillary action, on the grounds that the coefficients of interfacial tension are represented as forces (by unit length) acting on an immaterial line

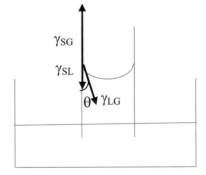

discussing the diagram (Viennot and Décamp 2018b). This issue of delayed critique was by far the most common problem among interviewees in the studies referred to above. In contrast, the opposite situation of 'early critique' proved to be rare, at least in relation to the selected topics (Viennot and Décamp 2018c).

2.2 Main Obstacles to Critical Analysis

The two cases of critical passivity—expert anaesthesia and delayed critique—can be ascribed to a number of issues (Viennot and Décamp 2018a). In both cases, most interviewees explicitly identified existing habits or 'teaching rituals' as an important obstacle to critique (Viennot 2006). In cases of delayed critique, interviewees also mentioned feelings of incompetence, which may relate to the content or to a broader inability to criticise a text.

In cases of 'expert anaesthesia', one likely (if speculative) obstacle is that where some logically crucial element is missing from the explanation, knowledge of the topic may lead to more or less unconscious completion. The same may be true in cases of inconsistent explanation—for instance, when presented with an 'isobaric hot air balloon', some experts may interpret the faulty hypothesis as 'approximately isobaric' without expressing any critique.

This example also points to a possible aggravating factor in critical passivity: the fact that, despite an invalid hypothesis, correct calculation (here, of Archimedes up-thrust) leads to a correct result (here, at first order). In a more recent investigation (Viennot 2019), this possible association between accurate calculation, correct result and fallacious modelling was explicitly articulated and further explored as 'misleading mathematical legitimacy' (MML). The goal was to explore the extent to which MML would prove difficult to detect because of the effect of mathematical pseudo-legitimation. To avoid simultaneous occurrence of two types of obstacle, the chosen case did not involve a teaching ritual. It concerns a solved exercise in first year at university. The task was to calculate the work done by the outside on an ideal gas experiencing an expansion in two distinct cases: where the expansion was either irreversible and non-quasistatic or reversible. Surprisingly, a correct calculation leading to a correct result used ideal gases relationship ($pV = nRT$, usual notations) to describe what happens *during* the irreversible transformation.

Box 1 A case of MML The irreversible expansion of an ideal gas in a solved exercise (Viennot 2019)

One mole of an ideal gas is situated in a cylinder, which is closed by a movable piston. It undergoes an expansion from pressure p_A to p_B ($p_A > p_B$) at constant temperature T. Calculate the work done on the gas in the two following cases:

a—irreversible* expansion, at constant external pressure ($p_{ext} = p_B$)

b—reversible expansion.

(a) Irreversible case

Work at T =constant for irreversible expansion of one mole of ideal gas: W_{irrev}

$$dW_{irrev} = -p_{ext}dV$$
$$dW_{irrev} = -p_B dV \text{ (because } p_{ext} = p_B)$$

For our system: $pV = NRT$ and $N = 1$ mol. Hence $V = RT/p$ and $dV = -RT\,dp/p^2$

$$dW_{irrev} = -p_B dV = RT\,p_B\,dp/p^2$$

$$W_{irrev} = \int_A^B RT p_B dp/p^2 = RT p_B \left[\frac{1}{p_A} - \frac{1}{p_B} \right] = p_B(V_A - V_B)$$

(b) ....

*The interviewer explains that this 'irreversible' transformation is also non-quasistatic.

In case a (irreversible and non-quasistatic), the intensive quantities (p and T) are not defined and use of the relationship of ideal gases is invalid. The interviewed beginning teachers (BTs) were uneasy about criticising such a text and expressed their surprise on realising the limits of validity regarding the relationship of ideal gases (even ideal gases are not always at equilibrium). Explaining their difficulty, most of the interviewees referred spontaneously to the fact that the calculation and result were correct.

3 Facilitating Critical Analysis in Teachers: Which Tools?

3.1 Criteria for Guiding Critical Analysis

These findings invite consideration of how teacher educators can activate critical analysis among their students. In this regard, metacognitive and affective factors seem to play an important (and possibly blocking) role. As one BT puts it, 'Who am I to criticise what important people have written?' It seems highly relevant, then, to convince students and teachers of their right—and, in many cases, their ability—to critically analyse such explanations. Expert anaesthesia's metacognitive component seems linked to the value of 'established' knowledge. Achieving some distance from habits is difficult for everyone, especially where the explanation in question seems to lead to an accurate conclusion. This confirms the need for educators to maintain a supportive and psychologically appropriate attitude during teacher preparation sessions.

As a more practical initial response to this challenge, teachers should be provided with analyses of contestable explanations according to specified criteria. In a recent book (Viennot and Décamp 2020), the authors proposed a twofold grid for this purpose. The first list of 6 items (with examples) identifies reasons that would indisputably refute the given explanation or that would at least introduce a strong element of doubt. For instance, the reader is invited to look for a possible internal contradiction or to pinpoint a missing logical link in an argument. The second list of 14 items specifies factors (also with examples) that might prompt misleading interpretations—for example, inappropriate designation of an entity or an over-selective diagram.

This tool facilitates the analytical approach to explanations used by the authors during teacher education sessions on critical thinking. As reported (Viennot and Décamp 2020), the participants simultaneously exhibited great interest, feelings of incompetence regarding this analytical process and mixed feelings about its impact on their classroom practice. They questioned what to do when an explanation they had previously considered appropriate for teaching purposes is found to be contestable for one reason or another. In this regard, the study of MML (Viennot 2019) revealed that an inconsistency in the solved exercise (Box 1) was not always enough to discourage participants from using it uncritically in class. For instance, one BT proposed to do so in order to 'have the laws used and a little bit of calculation and reinvest the laws in an exercise'.

3.2   *From Critical Analysis to Pedagogical Decision-Making*

These findings raise the question of why a teacher would choose to use one explanation rather than another despite its flaws. It seems that a grid that focuses discussion on an explanation's possible defects should also take account of potentially positive aspects such as simplicity or mnemonic power. In a pilot investigation of teachers' understandings of the current in a battery, eight of the eleven participants indicated a preference for an explanation that seemed consistent. However, seven of them selected a much less consistent explanation for teaching at higher education or university level on the basis of its greater simplicity, confirming the need for analytical grids to accommodate such decision criteria.

Such grids can lead to a 'quality diagnosis' for an explanation and can be used to improve that explanation. For instance, in the case of a hot air balloon, it can be said that internal and external pressure gradients are weak but different and are essential to flotation in the air. A grid can also inform a choice between several available explanations for a given phenomenon (Viennot 2020), allowing the teacher to refer to a kind of dashboard—that is, the set of 'quality diagnoses' of available explanations—in deciding what to do.

3.3 Avoiding Dogmatic Use of Grids

As shown by ongoing investigations, it is worth noting that the proposed analytical grids should be adapted for each example and context (including issues of available time) rather than simply being imported from a canonical source. As one obvious case in point, all the problems highlighted in (Viennot and Décamp 2020) regarding images have no relevance for explanations without images, reducing this to a useless box-ticking exercise. The relevance or irrelevance of other criteria (e.g. presence/absence of linear causal reasoning) is less obvious, and a binary response (yes or no) may be difficult for some criteria (e.g. explanation completeness). As Ogborn (Ogborn et al. 1996) argued, 'Explanations are like the tip of an iceberg, with a large amount of supporting knowledge lurking below the surface' (p. 65). On this view, any explanation could be characterised as incomplete because its prerequisites are not fully recalled. In the studies cited above, the focus is on logical incompleteness, but some difficulties are likely to arise in distinguishing these two cases of incompleteness. Moreover, both the logical completeness of an explanation and its generality can be undermined by a forgotten variable. The main virtue of using grids, then, is to foster critical reflection rather than ticking boxes in search of a 'right answer'. This issue is especially relevant for the criterion of simplicity, where a given teacher's judgement is strongly dependent on their past experience.

4 Concluding Remarks

Based on the above account, it seems clear that critical analysis of physics explanations remains an issue for teacher education. For most of the interviewed BTs, the proposed multi-criteria critical analysis or 'quality diagnosis' seemed an important step in improving an explanation or choosing one among several.

> It's a very good method in fact; first analyse, take all the different criteria, rank—yes, I didn't think at first—rank according to what we want to do. Clearly, for me, it is a very good method for analysis, if not the best possible.

Whatever the teacher's decision, preliminary critical analysis helps to clarify the advantages and disadvantages of their chosen explanation and facilitates discussion between colleagues.

> Well, it has the advantage once again of putting things in perspective and making what's in your head visible to everyone. In the context of a discussion, I think it is always important to communicate well with others.

That said, further research should explore the criterion of simplicity and its central importance in many teachers' decisions. Choosing an explanation for a given audience is likely to involve a trade-off between consistency and simplicity; critical capability can improve one's ability to evaluate an explanation's consistency, but simplicity poses at least two problems for teachers: how to judge what is 'simple' for students independent of one's own past experience and teaching habits, and to what extent consistency should be sacrificed if simplicity is deemed a priority. Further research on these two problems can help teachers to make more informed pedagogical decisions.

References

Décamp N, Viennot L (2015) Co-development of conceptual understanding, and critical attitude: Analysing texts on radiocarbon dating. Int J Sci Educ 37:2038–2063. https://doi.org/10.1080/095 00693.2015.1061720

Ogborn J, Kress G, Martins I, McGillicuddy K (1996) Explaining science in the classroom. Open UniversityPress, Buckingham

Viennot L (2006) Teaching rituals and students' intellectual satisfaction. Phys Edu 41:400–408. https://doi.org/10.1088/0031-9120/41/5/004

Viennot L (2019) Misleading mathematical legitimacy and critical passivity: discussing the irreversible expansion of an ideal gas. Eur J Phys 40:045701. https://doi.org/10.1088/1361-6404/ ab1d8b

Viennot L (2020) Educating to critical analysis in physics: Which direction to take? Phys Educ 55:015008. https://dx.doi.org/10.1088/1361-6552/ab4f64

Viennot L, Décamp N (2016a) Co-development of conceptual understanding and critical attitude: towards a systemic analysis of the survival blanket. Eur J Phys 37:015702. https://doi.org/10. 1088/0143-0807/37/1/015702

Viennot L, Décamp N (2016b) Conceptual and critical development in students teachers: first steps towards an integrated comprehension of osmosis. Int J Sci Educ 38:2197–2219. https://doi.org/ 10.1080/09500693.2016.1230793

Viennot N, Décamp L (2018a) Activation of a critical attitude in prospective teachers: from research investigations to guidelines for teacher education. Phys Rev Phys Edu Res 14:010133. https:// doi.org/10.1103/PhysRevPhysEducRes.14.010133

Viennot L, Décamp N (2018b) Concept and critique: two intertwined routes for intellectual development in science. In: Amin T, Levrini O (eds) Converging perspectives on conceptual change. Routledge, New York, pp 190–197

Viennot L, Décamp N (2018c) The transition towards critique: discussing capillary ascension with beginning teachers. Eur J Phys 39:045704. https://doi.org/10.1088/1361-6404/aab33f

Viennot L, Décamp N (2020) Developing critical thinking in physics. Springer (ESERA series), Dordrecht

Didactical Reconstructions in Knowledge Organization and Consolidation in Physics Teacher Education

Terhi Mäntylä

Abstract Physics teachers have an essential role in forming the attitudes and conceptions of future citizens towards science and technology, as well as in educating the future generations of scientists. Therefore, the physics teacher education must guarantee the best available education to pre-service physics teachers; sound knowledge of physics should be combined with a good understanding of the didactical and pedagogical aspects of teaching and learning. The situation is often that after university physics courses, the pre-service physics teachers' knowledge is still quite fragmented and incoherent. They also often lack the concept formation perspective to physics knowledge. I discuss here a research-based instructional approach that is developed for pre-service physics teachers for consolidating and organizing their subject matter content knowledge. In the core of the approach, graphical tools are called as didactical reconstructions of processes (DRoP) and structure (DRoS). The idea behind the reconstructions is that "new" physics knowledge is always constructed on the basis of previous knowledge. This leads to a network of quantities and laws, where the experiments and models construct the connections between the physics concepts. Finally, I discuss the implementation of didactical reconstructions in instruction and show that the didactical reconstructions help students to organize and consolidate their knowledge.

1 Introduction

Physics teachers have an essential role in forming the attitudes and conceptions of future citizens towards science and technology, as well as in educating the future generations of scientists. Therefore, the physics teacher education must guarantee the best available education to pre-service physics teachers. The backbone of physics teacher's expertise is the sound subject matter knowledge of physics. In addition, this

T. Mäntylä (✉)
Department of Teacher Education, University of Jyväskylä, Jyväskylä, Finland
e-mail: terhi.k.mantyla@jyu.fi

© The Editor(s) (if applicable) and The Author(s), under exclusive license to Springer
Nature Switzerland AG 2020
J. Guisasola and K. Zuza (eds.), *Research and Innovation in Physics Education: Two Sides of the Same Coin*, Challenges in Physics Education,
https://doi.org/10.1007/978-3-030-51182-1_7

should always be combined with a good understanding of the didactical and peda- gogical aspects of teaching and learning. However, often after university physics courses, the pre-service physics teachers' knowledge is still quite fragmented and incoherent (Bagno et al. 2000; Reif 1995, 2008). Physics teacher's subject matter knowledge has also its own requirements compared to the subject matter knowledge of physicists, especially understanding the process of physics knowledge formation, that are not usually addressed in pre-service physics teacher education. Therefore, the main challenge of pre-service physics teacher education is to provide opportu- nities and resources for pre-service physics teachers to (re)organize and consolidate their physics knowledge into larger, coherent and meaningful structures (Koponen et al. 2004). In order to meet the challenges discussed above, a teaching approach using didactical reconstructions of physics knowledge was developed. The didac- tical reconstructions and their implementations are introduced and discussed in detail in the previous research (see Mäntylä and Nousiainen 2014; Mäntylä 2012, 2013; Mäntylä and Hämäläinen 2015; Mäntylä and Koponen 2007; Koponen and Mäntylä 2006) and here, a concise overview of them is given.

2 Physics Teacher's Subject Matter Knowledge

Besides knowing or understanding the concepts of physics and being able to apply them in problem-solving, physics teacher must understand the origin of the concepts or be able to reconstruct it. Often the starting point is the law or the equation, where the concept (quantity) appears and the definition of the concept is the law or the symbolic relation of the concept to other concepts. Then, if student is able to solve problems using the concept, it is interpreted that student understands the concept. For a physics teacher, this is not enough. The physics teacher must understand, and how the law at first place was formed or can be formed. Similar idea of teacher's subject matter has been also been expressed by Shulman (Shulman 1986), when he discusses that the structures and organization of knowledge are part of the subject matter knowledge. This raises the epistemological perspective of knowledge formation in the centre of teacher's expertise.

Studies that examine expert's knowledge emphasizes that expert's knowledge is connected and organized around important concepts and ideas that guide thinking (cf. Reif 2008; Chi et al. 1981; Bransford et al. 2000). Likewise, Shulman discusses that teacher must know the essential and central topics of the discipline and also distinguish them from the less important concepts (Shulman 1986). This is desirable knowledge structure for a physics teacher too. However, how this kind of knowledge structure can be achieved (the epistemological perspective) is less discussed.

In summary, in addition to knowing (understanding) the concepts and facts of physics, a physics teacher must be able to answer the questions:

- How we know what we know?
- How the concepts and knowledge structures are formed or can be formed?

- How the knowledge relates to knowledge within the discipline (or knowledge of other disciplines)?
- What are the most important concepts of the discipline or a specific topic?

Physics teachers' organized and consolidated subject matter knowledge means that teachers' understand how physics concepts can be formed and how they are related to other concepts. This requires that the concepts are introduced in logical order, and in relation to each other, this also brings coherence in teaching or what is learned. The coherence also forms from the recurring knowledge forming processes. The main epistemological or methodological processes that form the concepts or can be used for forming the concepts are experimentation and modelling (Mäntylä and Nousiainen 2014; Mäntylä and Hämäläinen 2015; Koponen and Mäntylä 2006). Although the phenomenon of interest and the concepts change in different situations or topics, the procedure itself has recurring features. For instance, in the case of experiments, one can reconstruct a path from observations of qualitative laboratory experiments through qualitative experimentation and quantifying measurements to experimental laws (Mäntylä and Hämäläinen 2015; Mäntylä and Koponen 2007; Koponen and Mäntylä 2006). This path can be seen also in the didactical reconstruction of processes (DRoP), which is discussed below.

3 Didactical Reconstructions of Processes and Structures

The didactical reconstructions are reorganizations and simplifications of physics knowledge produced for the purpose of consolidate physics subject matter knowledge in a way that it enhances physics teacher's expertise. It means that besides taking into account the physics concepts, they emphasize the processes that (re)construct the concepts and further, support the (re)organization of physics knowledge structures.

The context of pre-service physics teacher education, for which the didactical reconstructions were developed, is such that the pre-service physics teachers have already studied the introductory and/or intermediate level university physics courses. The pre-service physics major teachers have usually studied the introductory and intermediate physics courses (around 70 cr), and the pre-service physics minor teachers have studied the introductory physics courses (around 25 cr).[1] Then they enter the physics teacher courses, which include such courses as concept formation of physics, school laboratory course for teachers and history and philosophy of physics. In the course "concept formation of physics", the didactical reconstructions are introduced to pre-service teachers. In Fig. 1, the development of physics teacher's subject matter knowledge is sketched concerning the didactical reconstructions. As discussed earlier, the pre-service teachers still have a fragmented view of physics: they know different concepts, definitions and laws, but the bits and pieces do not form coherent knowledge structures. They are also unable to explain or justify how

---

[1]The pre-service physics major teachers study physics altogether 130–140 cr and the pre-service physics minor teachers at least 60 cr.

Fig. 1 Transformations of subject matter knowledge concerning didactical reconstructions

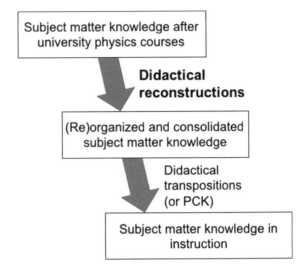

a specific concept or law is obtained or can be obtained. The didactical reconstructions are aimed for correcting the situation. A solid and organized view of physics knowledge is needed in teacher's profession in order to plan physics instruction and to adjust the level suitable for the pupils or students. This planning and adjusting of subject matter knowledge are addressed by didactical transpositions (Chevallard and Bosch 2014) or pedagogical content knowledge (PCK) (Shulman 1986)

The didactical reconstructions are based on the generative knowledge justification, which means shortly that experimentality and modelling are the central procedures or methodologies that construct the physics knowledge and they do it in an intertwined way (Mäntylä and Nousiainen 2014; Koponen and Mäntylä 2006). For practical purposes, the didactical reconstructions are presented in a visual form of flowcharts and concept networks, because the visual representations have proved to be effective in supporting the construction of knowledge structures (e.g. Bagno et al. 2000; Van Heuvelen 1991).

When the didactical reconstructions are applied to a certain topic, an analysis of content structure is done. Although the didactical reconstructions are not intended to be applied in actual school teaching as such, the important ideas of the topic from the perspective of the future profession as a teacher is kept in mind as Duit, Gropengießer and Kattmann has discussed in the first step of their Model of Educational Reconstruction (Duit et al. 2005). Also, the cognitive—historical analysis (Nersessian 1992) has been applied didactical way in order to preserve a kind of authenticity of the knowledge formation process (Mäntylä 2013). Next, the didactical reconstruction of processes (DRoP) is introduced first and the didactical reconstruction of structures (DRoS) after it, because the processes create the structures.

3.1   Didactical Reconstruction of Processes: DRoP

Didactical reconstruction of processes represents the knowledge generation process based on the interplay of experiments and models. The process is simplified into eight steps of knowledge construction. These steps are schematically summarized in the flowchart shown in Fig. 2. The eight steps are (Mäntylä and Nousiainen 2014; Mäntylä 2012):

1. *Observation and identification of phenomenon.* A phenomenon is identified through qualitative laboratory experimentation, which is guided by already known theory through models.
2. *Qualitative experimentation.* Experimentation helps to observe and find the changing and constant properties (or qualities).
3. *Qualitative dependency.* The result of the qualitative experimentation is a qualitative dependency, which forms the basis for designing quantifying measurement.
4. *Model system and measurement.* Quantifying measurements are designed based on the qualitative dependencies and the model for measurements. Experiment is modelled.
5. *Representation.* The measurement results are represented in a graph.
6. *Experimental law and model representation.* The new experimental law is justified and interpreted in the light of earlier knowledge, i.e. theory, through modelling.
7. *Extension of theory.* The new tentative law is annexed to existing theory through generalization.
8. *Interpretations and predictions.* The law is tested in different situations in order to validate it.

Fig. 2 Schematic representation of DRoP (6, p 795)

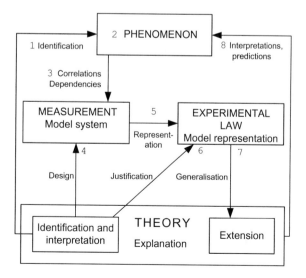

In practice, the DRoP serves as a graphical tool, which helps pre-service physics teachers to recognize the important processes and structural features of knowledge construction and learn to use these features to give ordered form for their learning and also teaching (Mäntylä and Nousiainen 2014). After applying the DRoP to a certain topic, the pre-service teachers should be able to answer the questions: "How we know what we know"? and "How the concepts are formed or can be formed"? In practice, the pre-service teachers are filling the boxes and the links at the general level and they supplement it with a detailed description of the steps, for an example of this; see (Mäntylä 2013).

3.2 Didactical Reconstruction of Structures: DRoS

In DRoP, "new" physics knowledge is always constructed on the basis of previous knowledge. When the process is repeated or applied to new phenomena, it leads to a network of quantities and laws, where the experiments and models construct the connections (laws) between the physics concepts. This network has a structure and an order, in other words, a hierarchy. The emerging concept network is the DRoS, and the forming process of DRoS is tried to capture in Fig. 3.

The nodes of the network are quantities (presented in rectangular shapes) and sometimes phenomena (rounded corners) and through DRoP (hexagonal shape) the quantities are related to forming laws (oval shapes). Often, when the quantities are related in the form of law, a new quantity is established. For example, the relation of electric current and voltage forms Ohm's law, and at the same, a new quantity, resistance, is established (Mäntylä and Hämäläinen 2015). The main features of the DRoS are (Mäntylä and Nousiainen 2014):

- It is an ordered node-link representation that includes the major quantities and laws of a certain topic. Ordering emerges, because a new quantity or law is constructed based on previous quantities or laws.
- The DRoP defines which concepts are connected, and it gives the direction to the link.
- The different kinds of concepts are distinguished from each other using different node shapes.

In practice, instead of going through the DRoP with its eight steps, the most relevant (quantifying) experiment or modelling procedure is described. In instruction, when pre-service teachers are constructing the concept networks, they describe the nodes in a separate supplement. After applying the DRoP to a certain topic, the pre-service teachers should be able to answer the questions:

- How the knowledge structures are formed or can be formed? Constructing the concept network and its concise form as node-link-representation forces pre-service teachers to think about the forming of the knowledge structure

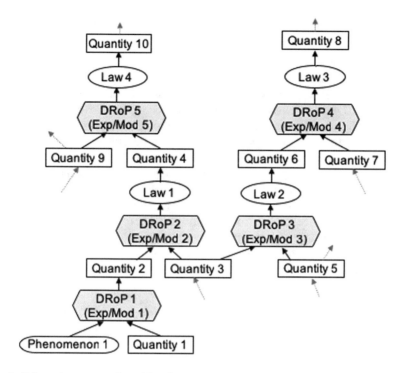

Fig. 3 Schematic representation of DRoS

- How the knowledge relates to knowledge within the discipline? Although the experimentation and modelling concerns different phenomena and concepts, they have recurring features and processes, so there is procedural or methodological coherence within the different topics of the discipline. The concepts' meanings are also augmented within physics; for instance, the force is first introduced in mechanics, and later, it is applied in electromagnetism.
- What are the most important concepts of the discipline? When the constructed concept networks are examined, it can be inferred, that concepts, which have a high amount of links are central, and concepts with only one link are probably peripheral.

The DRoS makes possible to examine different topics at various depth and range. It can be used for examining the development of a specific concept such as in the case on temperature (Mäntylä and Koponen 2007) or it can be used for examining a large network of quantities and laws such as in the case of electromagnetism (Mäntylä and Nousiainen 2014; Nousiainen 2013).

4 Didactical Reconstructions in Practice: Summary of Evidence from Case Studies

The basic implementation of the didactical reconstructions in the instruction of pre-service physics teachers is presented in Fig. 4. The instruction starts with introducing the DRoP and DRoS and discussing the ideas behind them, such as concept formation process in physics. Then pre-service teachers apply the tools in a certain topic or context. In instruction, the topic is discussed at a general level, and pre-service teachers have to adjust and apply the information from instruction to their flowcharts or concept networks. The pre-service teachers also get feedback from their peers and instructor about their initial flowcharts or concept networks. On the basis of these, the pre-service teachers revise and finalize their initial flowcharts or concept networks. Usually, the pre-service teachers construct the charts or networks in pairs or in small groups, so that they have a lot of opportunities to discuss and reflect their knowledge. For research purposes, individually done charts or networks with their supplements have been collected.

4.1 DRoP in Case of Electromagnetic Induction Law

The DRoP has been applied in the case of (re)constructing the electromagnetic induction law (Mäntylä 2012, 2013). The data consisted pre-service teachers' initial and final flowcharts and the written supplements explaining the flowcharts. Part of the pre-service physics teachers was also interviewed. The results of the analyses of the initial reports showed that although the pre-service teachers had already studied the topic of electromagnetic induction, it was poorly understood and the descriptions of the forming of the induction law were incoherent and poorly justified. In final reports, in most of the cases, the justifications were improved. Also, the use of experiments and models in the justifications improved greatly and, in many reports, there was a recognizable path from qualitative laboratory experiments of induction current to

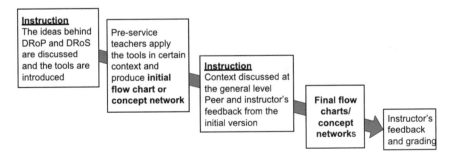

Fig. 4 Implementation of the didactical reconstructions in instruction

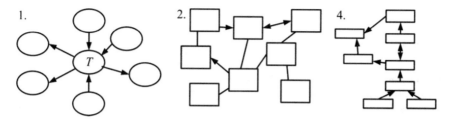

Fig. 5 Schematic structures of pre-service physics teachers' concept networks. 1. Centralized, 2. Fragmented, 3. Mixed (1 or 2 and 4) and 4. Hierarchical

the electromagnetic induction law. It has to be noted that the final products were not perfect and often, there still was room for improvement. However, most of the times, there was a clear development from initial to final reports.

4.2 DRoS in Case of Temperature

In case of temperature, the DRoS was applied to examine the quantitative development temperature and how new ways of measure (define) temperature augments its meaning (Mäntylä 2013). The data consisted pre-service teachers' initial and final concept networks; in addition, few pre-service teachers were interviewed. In Fig. 5, the schematic structures classified from pre-service teachers' concept networks are shown. Initial network structures were mainly centralized or fragmented, which means, that there was no recognizable development in temperature concept. Most of the final network structures were hierarchical, and thus, there was a developmental path from sensory experience and measuring the temperature from thermal expansion of liquids to measuring temperature with help of gas laws and defining the absolute temperature. The quality of concepts in the concept networks also improved. The study showed that the idea of the evolving meaning of a concept was at first, unfamiliar to pre-service teachers; the concepts just exist. The DRoS in case of temperature helped pre-service physics teachers to understand the progressive nature of physics concepts and the role of experiments and models in it.

4.3 DRoS in Case of the Network of Electromagnetism

The idea of DRoS has also been applied to capture the order and relations of electromagnetism concepts (Mäntylä and Nousiainen 2014; Nousiainen 2013). The data consisted pre-service physics teachers' concept networks of electromagnetism. The concept networks were classified into three different structures: webs (richly connected), necklaces (loosely connected) and chains (poorly connected). Most of the networks were either loosely or poorly connected and expert-like web structures formed only one-fifth of the structures. The analysis of pre-service physics teachers' explanations and justifications of the relations of the concepts also showed that pre-service teachers struggled in providing sound explanations and justifications. In this case, the knowledge structure to be covered was larger, which could explain that the development and consolidation in knowledge were more moderate than in the cases discussed above.

5   Discussion and Conclusions

The didactical reconstructions and their visual formulations function as metacognitive tools that forces pre-service teachers to think physics and its concepts from the perspective of what is it about and how the concepts are formed or used in physics concept formation. This goes beyond typical textbook definitions based on equations (laws). At the end, the laws can be treated in deterministic way and it improves the applicability of physics knowledge. However, in the beginning, when the learning process of physics concepts is still on, it is essential to approach the concepts and their relations in causal way and starting from the phenomena that they relate to. The didactical reconstructions provide order and hierarchy to physics concepts that can be utilized in teaching physics. However, it has to noted, that there is not just one right way of organizing the concepts, instead there are several good ways to do that. The didactical reconstructions help to find to answers to the subject matter knowledge questions introduced earlier and the pre-service physics teachers also appreciate them:

> This is useful in a way that it is important for a teacher to perceive these big pictures and a kind of hierarchy of laws and their relations...and a sort of order in knowledge construction. (Mäntylä 2012)

> Learning physics is like climbing upwards step by step, and every step is needed. This is useful [making concept networks] because it organizes thinking, and one easily recognizes in what step something is missing. It is possible to build a whole structure of what one has learned (Mäntylä and Koponen 2007).

The results show that the opportunities and resources invested in these case studies of applying the didactical reconstructions improve the pre-service physics teachers understanding on physics subject matter knowledge of the topics covered in pre-service teacher education. However, there is no time or resources to cover all topics. Yet, the feedback from pre-service teachers encourages us to think that the time invested in few carefully chosen topics helps pre-service physics teachers to develop their thinking from fragmented collections of definitions towards more coherent knowledge structures. The pre-service teachers have also learned more deeply the topics covered using the didactical reconstructions; besides the factual and definitional knowledge, they have learned to (re)construct their knowledge and reflect the forming knowledge structures. The pre-service teachers have learned to understand the role of experiments and models in knowledge production and they have improved in their knowledge justification. In short, the didactical reconstructions have helped pre-service physics teachers to consolidate and (re)organize their subject matter knowledge.

References

Bagno E, Eylon B, Ganiel U (2000) Am J Phys Suppl 68:S16
Bransford JD, Brown AL, Cocking RC (eds) (2000) How people learn: brain, mind, experience, and school. National Academy Press, Washington, DC
Chevallard Y, Bosch M (2014) Didactic transposition in mathematics education. In: Lerman S (ed) Encyclopedia of mathematics education. Springer, Dordrecht
Chi MT, Feltovich PJ, Glaser R (1981) Cog Sci 5:121
Duit R, Gropengießer H, Kattmann U (2005) Towards Science education that is relevant for improving practice: the model of educational reconstruction developing. In: Fischer H (ed) Standards in research on science education. Taylor & Francis, Leyden, pp 1–9
Koponen I, Mäntylä T (2006) Sci Ed 15:31
Koponen I, Mäntylä T, Lavonen J (2004) Eur J Phys 25:645
Mäntylä T (2012) Res in Sci Edu 42:791
Mäntylä T (2013) Sci Edu 22:1361
Mäntylä T, Hämäläinen A (2015) Sci Ed 24:699
Mäntylä T, Koponen I (2007) Sci Ed 16:291
Mäntylä T, Nousiainen M (2014) Sci Edu 23:1583
Nersessian NJ (1992) How do scientists think? capturing the dynamics of conceptual change in science. In: Giere RN (ed) Cognitive models of science. University of Minnesota Press, Minneapolis, MN, pp 3–45
Nousiainen M (2013) Sci Ed 22:505
Reif F (1995) Am J Phys 63:17
Reif F (2008) Applying cognitive science to education: thinking and learning in scientific and other complex domains. The MIT Press, London
Shulman LS (1986) Ed Res 15:4
Van Heuvelen A (1991) Am J Phys 59:891

Examining Students Reasoning in Physics Through the Lens of the Dual Process Theories of Reasoning: The Context of Forces and Newton's Laws

Mila Kryjevskaia and Nathaniel Grosz

Abstract Prior research identified a common phenomenon observed in introductory physics courses (and beyond): Students often demonstrate competent reasoning on one task, but not on another, closely related task. Sometimes, students simply do not possess the formal knowledge necessary to reason productively (referred to as mindware). In other cases, students seem to abandon the formal reasoning in favor of more appealing intuitive ideas. These observed inconsistencies can be accounted for by the dual process theories of reasoning, which assert that cognition relies on two thinking processes. The first process is fast, intuitive, and automatic; the second is slow, rule-based, and effortful. The tendency toward mediating automatic responses via productive engagement of the slow and analytic process is called the cognitive reflection skills. We present results from an empirical investigation suggesting that both mindware and cognitive reflection skills play key roles in physics performance. Moreover, even in the presence of mindware, students with low cognitive reflection skills tend to reason intuitively on certain types of physics tasks. We argue that efforts directed toward the development of instructional interventions that take into account tendencies in student reasoning are critical for achieving further improvements in physics performance.

1 Introduction

Experienced physics instructors are familiar with student responses that reveal reasoning inconsistencies in many contexts. These inconsistencies stem from a variety of factors. Students who did not develop an appropriate level of conceptual understanding during instruction are more likely to reason incorrectly and inconsistently. It is also common for novice learners to excessively rely on their everyday

M. Kryjevskaia (✉) · N. Grosz
Department of Physics, North Dakota State University, 1211 Albrecht Blvd, Fargo, ND, USA
e-mail: mila.kryjevskaia@ndsu.edu

© The Editor(s) (if applicable) and The Author(s), under exclusive license to Springer
Nature Switzerland AG 2020
J. Guisasola and K. Zuza (eds.), *Research and Innovation in Physics Education: Two Sides of the Same Coin*, Challenges in Physics Education,
https://doi.org/10.1007/978-3-030-51182-1_8

perceptions of "how things work" instead of reasoning based on physics principles (Singh 2002; diSessa 1993; Sabella and Cochran 2004; Lising and Elby 2005; Lindsey et al. 2018). The purpose of this study is to examine inconsistencies in student reasoning that arise even in the presence of a robust conceptual understanding. Prior research on student reasoning revealed that even students who demonstrate that they possess the relevant knowledge and skills necessary to answer many types of physics questions correctly often fail to do so (Kryjevskaia 2019; Kryjevskaia et al. 2014, 2015). It appears that these students apply the acquired knowledge and skills in a selective manner. On the questions that require a straight-forward application of physics knowledge, students are able to reason correctly. However, on many questions that require the same application of physics principles, but also tend to elicit intuitively appealing (but incorrect) responses, those same students appear to either (1) abandon the correct reasoning approaches in favor of intuitive ideas or (2) apply formal knowledge in an incorrect manner in attempt to confirm what they already erroneously believe to be a correct response.

This investigation was conducted in an introductory calculus-based mechanics course for science and engineering majors. We examined student performance in the context of forces and Newton's laws of motion. Section 2 of this paper focuses on applying the dual-process theories of reasoning developed in cognitive psychology in order to interpret inconsistencies in student reasoning. Section 3 centers around the identification of factors and instructional circumstances that may impact student reasoning pathways and lead to inconsistences. In Sect. 2, we first introduce a pair of screening and target questions that highlight inconsistencies in reasoning and allow (to the extent possible) for the disentanglement of student conceptual understanding from their reasoning approaches. Then, we apply the dual-process theories of reasoning (DPToR) in order to understand, in a mechanistic fashion, reasoning paths that lead to inconsistencies even in the presence of a relevant conceptual understanding. The results are then used to develop a sequence of instructional interventions to address the observed inconsistencies. Finally, a set of assessment tasks aimed at probing the effectiveness of these interventions is introduced and the results are discussed. Section 3 focuses on an empirical investigation of relationships among (1) student conceptual understanding, (2) cognitive reflection skills, which allow reasoners to recognize and override intuitive thoughts by applying formal reasoning approaches, and (3) student performance on physics tasks that tend to elicit persistent and incorrect intuitive responses.

2   Applying Dual Process Theories of Reasoning to Interpret and Address Reasoning Inconsistencies

This section focuses on identifying and interpreting inconsistencies in student reasoning. The results are used to develop a set of instructional interventions and to assess their effectiveness.

2.1   Screening and Target Questions

In order to pinpoint the nature of student incorrect responses that appear to persist even after instruction, it is necessary to disentangle student conceptual understanding from reasoning approaches. To do so, we have been applying the *screening–target question* methodology. *Screening questions* are designed to give students an opportunity to demonstrate whether or not they acquired basic physics knowledge during instruction and can apply it correctly in situations that are challenging but not known for eliciting incorrect intuitive ideas. *Target questions*, on the other hand, call for the application of the same set of physics ideas and accompanying reasoning approaches; however, target questions tend to elicit persistent intuitively appealing, but incorrect, responses. We then examine the reasoning on target questions of those students who answer screening questions correctly. This approach ensures that persistent incorrect responses to target questions do not stem from the lack of relevant physics knowledge but appear to be indicative of reasoning difficulties. An example of a pair of screening and target questions in the context of forces and Newton's laws is shown in Fig. 1.

Question 1 (referred to as *the Block question* in this paper) serves to screen for the presence of conceptual understanding. It involves a heavy block at rest on a table with a massless rod glued to the block, as shown in Fig. 1. Students are told that the weight of the block is 50 N, and the force exerted on the block by the table is 30 N. They are asked to determine whether the force that the rod exerts on the block is upward, downward, or zero. Students are expected to draw a free-body diagram of the block and apply Newton's second law: Since the object is at rest, the net force

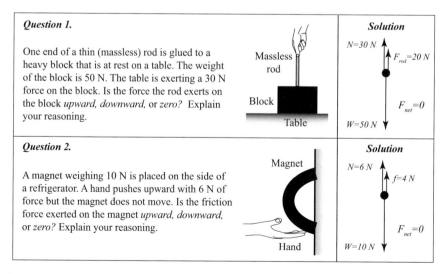

Fig. 1 Pair of screening and target questions

on the object must be zero; as such, the force of the rod on the block, F_{rod}, must be 20 N in the upward direction.

Question 2 (referred to as *the Magnet*) is a target question. It requires the same reasoning but tends to elicit responses inconsistent with the correct approach. Specifically, in this question, students consider a magnet placed on a refrigerator door. They are told that the magnet weighs 10 N and that a hand pushes upward with a 6 N force, but the magnet remains at rest. Students are asked to determine whether the force of friction between the magnet and the refrigerator door is upward, downward, or zero. Physics experts may regard the target question as being nearly identical to the screening since both require the same reasoning steps, as shown in Fig. 1. However, the data show that, while ~71% of students answered the screening question correctly, only ~23% applied the same line of reasoning on the target question. Moreover, only about a third (31%) of the students who answered the screening question correctly were able to provide correct answers to the target question with correct reasoning. The rest of the students argued that *"The friction opposes the applied force by the hand and therefore must point in the downward direction."* In order to interpret the observed inconsistencies in student reasoning, we applied the dual-process theories of Reasoning described in detail in Sect. 2.2.

2.2 Theoretical Framework: Dual Process Dual-process of Reasoning, Mindware, and Cognitive Reflection

According to the DPToR, two distinct processes are involved in most reasoning tasks (Kahneman 2011; Evans 2006). Process 1, often referred to as the *heuristic process,* is fast, automatic, and subconscious. Process 2, referred to as the *analytic process,* is slow, deliberate, and effortful. The interaction between the two processes is illustrated by the diagram in Fig. 2. Once a reasoner becomes familiar with a presented situation, the heuristic process immediately and subconsciously suggests a mental model of (or a way of thinking about) this situation based on prior knowledge and experiences,

Fig. 2 Diagram illustrating interactions between the heuristic and analytic processes (Evans 2006)

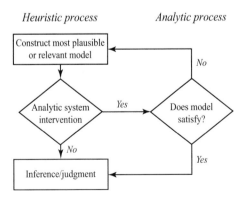

contextual cues, and plausibility. In everyday life, the first available mental model is often referred to as intuition or "gut feelings." A critical aspect of the DPToR is that reasoners view the world around them through the lens of the heuristic process, which cannot be turned off. This implies that intuition plays a vital role in reasoning because it provides an entry point into a reasoning path. That is, the intuition-based response is cued before the slow and rule-based analytic process has an opportunity to intervene. If a reasoner feels confident in the first intuitive mental model, the analytic process is often entirely bypassed (Thompson et al. 2011; Thompson et al. 2013). This direct path from the first available mental model to the final inference is often called "cognitive miserliness (Johnson-Laird 2006; Toplak et al. 2011a)."

It is important to note that an engagement of the analytic process does not necessarily guarantee a productive analysis of the first available response. In many cases, reasoners do not spontaneously look for evidence that may refute what they strongly believe to be true. Instead, they often exhibit the opposite behavior. Many reasoners tend to look for evidence that supports their initial and intuitively appealing responses (Nickerson 1998). (This reasoning phenomenon is known as a "confirmation bias.") Therefore, even if the analytic process is engaged, an intuition-based incorrect response may persist. Only in the presence of a strong "red flag" may the analytic process be placed on alert, which may lead to a feeling of dissatisfaction and subsequent rejection of the first available mental model. If this occurs, then the analytic process shifts back to the heuristic process and the reasoning cycle repeats (i.e., a new mental model is suggested, which may be further scrutinized by the analytic process).

Since the first available mental model serves as an entry point into a reasoning path, it is logical to hypothesize that those reasoners who are better at recognizing instances of their own unsupported intuitive thoughts and who are able to engage in an unbiased evaluation of such thoughts would be more likely to reason successfully. The ability to mediate intuitive thinking by reasoning more analytically is called *cognitive reflection*. The cognitive reflection test (CRT), developed by Frederick, has been used to measure this ability (Frederick 2005) (see Fig. 3). There is a general consensus among researchers that the CRT is not a measure of cognitive or mathematical abilities, but it is "a particularly potent measure of the tendency toward miserly processing (Nickerson 1998; Frederick 2005; Toplak et al. 2011a, b; Stanovich 2009; Pennycook et al. 2016; Campitelli and Gerrans 2014)."

A distinct characteristic of each CRT question is that it tends to immediately elicit an intuitively appealing, but incorrect, response. For example, most people give "10 cents" as an answer to the first CRT question. While this answer may appear highly

---

1. A bat and a ball cost $1.10 in total. The bat costs $1.00 more than the ball. How much does the ball cost?
2. If it takes 5 machines 5 minutes to make 5 widgets, how long would it take 100 machines to make 100 widgets?
3. In a lake, there is a patch of lily pads. Every day, the patch doubles in size. If it takes 48 days for the patch to cover the entire lake, how long would it take for the patch to cover half the lake?

Fig. 3 Three-item cognitive reflection test

plausible, it is incorrect. Upon brief examination, the majority of the reasoners are able to recognize that if the ball cost 10 cents, then the ball and the bat together would cost $1.20. As such, the intuitively appealing response of "10 cents" must be incorrect. Subsequently, most reasoners are able to deduce, without gaining any additional knowledge of arithmetic, that the correct answer is 5 cents. (The correct answers to the CRT questions are 5 cents, 5 min, and 47 days, respectively.) This example illustrates the nature of human reasoning: Even in the presence of relevant knowledge (e.g., basic rules for addition and subtraction), many reasoners tend to accept intuitively appealing responses as correct without giving them any additional thought. As such, an incorrect response to a question that elicits a strong intuitively appealing response may not necessarily reflect a lack of relevant knowledge. Instead, it may be indicative of the reasoner's tendency toward cognitive miserliness. Indeed, Stanovich suggests that learners may reason incorrectly because either or both of the following occur: They lack an appropriate knowledge of the rules and concepts required to answer correctly (referred to as "*mindware*"); they behave as cognitive misers (Stanovich 2009).

We argue that the DPToR, together with the accompanying constructs of mindware and cognitive reflection, may be used to interpret the pattern of inconsistent responses on the screening–target pair above. We argue that the student performance on the screening question suggests that ~71% of the students have acquired the knowledge and skills necessary to answer the target question correctly. However, the majority of these students seemed to abandon the correct line of reasoning in favor of a more readily available mental model consistent with the notion that "a force of friction opposes the applied force." This mental model may be intuitively appealing because it is often used correctly and appropriately in a variety of situations discussed in introductory mechanics courses. For example, a block at rest on a horizontal surface pushed by a single horizontal force does experience friction in the direction opposite to the applied force. As such, it is not surprising that many students immediately and confidently accepted this response as "correct," causing any further engagement of the analytic process to seem unnecessary. This strong "feeling of rightness" may have also prevented the students from recognizing a glaring red flag: The intuitive response does not take into account the ubiquitous and ever-present gravitational force.

It is important to note that in the context of the DPToR, intuition is defined as the first available mental model suggested by the heuristic process. Therefore, intuition may not necessarily be based on everyday knowledge and experiences. In fact, it may be rooted in formal knowledge as well. Based on the work by Simon, Kahneman views intuition "as nothing more and nothing less than recognition (Kahneman 2011; Simon 1992)." This explains why, in a given domain, experts' intuition is distinctly different from that of novices. Experts rely on a vast repertoire of prior experiences, which allows them to quickly recognize a presented situation with a high level of accuracy. Novices' intuition, on the other hand, is much less expansive and therefore less reliable. Novices are more likely to recognize a situation erroneously. For example, in this study, a novice learner may incorrectly perceive the target question as being "about friction" and, therefore, may think that "ideas related to friction must be used."

2.3 Implications for Instruction and Development of Instructional Intervention

While the DPToR are valuable for interpreting inconsistencies in reasoning, they do not provide guidance for designing instructional interventions aimed at developing the skills necessary for a productive engagement of the analytic process. Moreover, while significant efforts in physics education research (PER) have been directed toward strategies that promote the development of conceptual understanding and the accompanying reasoning, research that takes into account student intuition is still emerging. As such, the design of our instructional interventions was motivated by a variety of ideas, some of which are grounded in prior research, while others are exploratory. Below we briefly discuss our motivations for the specific instructional strategies implemented in this study.

- Prior research suggests that in order to reason productively, it is necessary to possess both strong conceptual understanding (mindware) and the ability to mediate intuitive responses by reasoning more analytically (cognitive reflection) (Stanovich 2009). In addition, in contexts that tend to elicit intuitively appealing but erroneous responses, refinements of instructional materials solely focused on strengthening relevant conceptual understanding are not likely to generate desirable improvements in student performance (Kryjevskaia et al. 2012). Since the majority of the students in this study appeared to already possess the necessary mindware, the instructional efforts were directed toward helping these students recognize inconsistencies in their reasoning. We argue that this type of instruction could provide opportunities for students to learn how to recognize instances of intuitive thoughts, check such thoughts for validity, and replace them (if necessary) with a carefully justified response.
- Activities that foster spontaneous, socially mediated metacognition (such as group work) tend to minimize the occurrences of reasoning inconsistencies (Goos et al. 2002). Hence, our instructional interventions included both individual and group work.
- We speculated that, depending on the level of conceptual understanding and cognitive reflection skills, some students may only require a quick prompt to recognize and override intuitive ideas. Others, however, may need a more involved instructional approach. As such, our intervention included three stages with different levels of guidance, as described below.

All activities were implemented in a web-based format and were administered upon completion of the regular instruction related to forces and Newton's laws (i.e.,

Table 1 Student performance on the screening (the block) and target (the magnet) questions

	Performance on screening question (%)	Performance on target question			
		Pre-intervention (%)	Stage 1 (%)	Stage 2 (%)	Stage 3 (%)
All students	71	23	34	50	74
Correct on screening question	100	31	43	54	77

lectures, labs, and homework assignments). A set of assessment questions was given on a course exam in a traditional paper-based format. All activities and assessments required students to provide answers and explain their reasoning.

2.3.1 Pre-intervention Stage Involving Individual Work: Screening–Target Pair

The pre-intervention stage was designed to give students an opportunity to consider the screening–target pair on their own, outside of class, before instructional interventions developed as part of this project. The row of Table 1 labeled "*All students*" shows the percentages of correct responses of all the students participating in the study ($N = 76$). The bottom row illustrates the performance of those students who answered the screening question correctly ($N = 54$).

2.3.2 Stage 1 Involving Individual Work: Intervention Designed to Raise Awareness of Similarities Between the Screening and Target Questions

Stage 1 was a part of the individual work completed outside of class. It immediately followed the pre-intervention stage discussed above. Students considered a provided (correct) solution to the Block question and were asked to indicate whether or not they agree with the solution. As shown in Fig. 4, the solution was intentionally

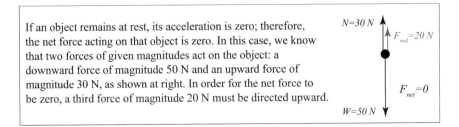

Fig. 4 Feature-free solution to the block question

designed in a feature-free form: It did not reference the block or mention the names of the forces acting on it. Instead, it was meant to evoke general reasoning steps necessary for a correct analysis of forces acting on an object at rest. This stage was intended to function as a quick prompt. It was not designed to provide a direct link between the screening and target questions. However, it was expected to trigger a more productive engagement of the analytic process by raising awareness of the similarities between the two situations and by prompting recognition of red flags in students' original approaches to the Magnet question. In order to examine whether or not this intervention functioned as intended, students were asked to consider the Magnet question again and to modify their solutions, if necessary. Data revealed that the percentage of correct responses to the Magnet question after Stage 2 increased by ~10%, as shown in Table 1. The result suggests that for the majority of the students this intervention did not provide enough guidance to start questioning the validity of their intuitively appealing original responses to the Magnet question.

2.3.3   Stage 2 Involving Group Work: Screening and Target Pair

This stage was implemented in a laboratory a few days after the individual work was completed. Students worked with their regular lab partners in groups of two or three. They considered the same pair of screening and target questions and were instructed to discuss their answers until a group consensus was reached. Much like in the individual stages, students submitted their responses in the web-based format. However, each group recorded a consensus response only. After Stage 2, more than half of the students provided (or at least agreed with) the correct reasoning to the Magnet question. While the success rate on the Magnet question after the group work is much higher than that after the first individual attempt, the overall low fraction of correct responses illustrates a highly persistent nature of intuitively appealing responses. Even those students who demonstrated that they possess the knowledge and skills required to reason correctly do not appear to be able to recognize the red flags in their incorrect reasoning approaches to the Magnet question. They also do not appear to see the need to check the validity of their responses by attempting to apply an alternative solution, such as one based on Newton's second law.

2.3.4   Stage 3 Involving Group Work: Sequence of Guiding Questions

Groups that did not answer the Magnet question correctly after Stage 2 were directed to a final sequence of questions designed to provide more involved, step-by-step guidance. The goals of Stage 3 were two-fold: (1) motivate the applicability of Newton's second law to the context of the Magnet and (2) refine the notion of "friction opposing the applied force." Specifically, students were presented with two situations involving a magnet at rest on a refrigerator door. In the first situation, a hand did not interact with the magnet. The second situation was identical to the original Magnet question in which the hand was applying a 6 N force upward. Students were asked to draw free-body diagrams and determine the net force on the magnet in each case. Then, the students were prompted to consider what their answers suggest about the direction of the force of friction. The majority of students correctly concluded that, in the first situation, the friction opposes the force of gravity and, therefore, must point upward. However, only a fraction of the students was able to give the correct answer to the second situation as well. A significant number of the students continued to argue that the friction opposes the applied force by the hand, thus entirely neglecting the effect of gravity (see Table 1).

2.3.5   Instructional Interventions: Summary of Results and Further Assessment

Each stage in the sequence of interventions yielded a modest improvement in student performance. These results reveal a significant impact of intuition on reasoning. Intuitively appealing, but incorrect, responses appear to be difficult to dislodge even in the presence of mindware. After multiple instructional attempts, many students still do not appear to see red flags in their reasoning. They also have not yet developed the habit of searching for alternatives, which could potentially lead to identification and resolution of reasoning inconsistencies.

After the sequence of instructional interventions, only ~75% of all students provided (or seemed to agree with) the correct solution to the Magnet question. To probe the robustness of the correct reasoning, a set of assessment questions was administered on a regular course exam approximately two weeks after the interventions. The assessment tasks were also designed in the context of a magnet at rest on a refrigerator door. We recognized that the familiarity of this context in itself could serve as a red flag for those students who had struggled with the question during interventions. According to the DPToR, the state of a heightened alert stimulated by this context could lead to a more productive engagement of the analytic process and, subsequently, could result in a higher success rate on the assessment tasks. To ensure that the exam tasks contained some elements of novelty and discouraged memorized responses, students were asked to consider four different cases with a magnet weighing 5 N at rest on a refrigerator door. In Case 1, a string pulls on the magnet upward with 3 N of tension. In Case 2, a hand pushes upward with a 5 N force. In Case 3, a string pulls downward with 3 N of tension. In Case 4, a stack of coins that weighed 3 N sits on the top of the magnet. In all cases, students were asked to determine the direction and the magnitude of the force of friction between the magnet and the refrigerator. Responses that contained correct answers with correct reasoning in all four cases were counted as correct. Results suggest that, despite the familiarity of the context, ~30% of the students who answered the Magnet question correctly at some point during interventions provided intuition-based incorrect responses based on the notion that "friction opposes the applied force" (e.g., friction points in the direction opposite to the force of tension in Cases 1 and 3).

3 Identification of Factors and Instructional Circumstance that Lead to Inconsistencies in Student Reasoning

Conceptual understanding is often necessary but not sufficient to reason correctly. The results in Sect. 2 suggest that even with necessary mindware, students often tend to abandon correct reasoning in favor of more compelling intuitive ideas. Moreover, even those students who reasoned correctly previously (or appeared to agree with a correct consensus argument) still may give intuition-based responses on a nearly identical assessment task a short time later. The DPToR assert that, in addition to mindware, cognitive reflection skills are critical for mediating intuitive responses. As such, an overarching goal of this part of the study is to conduct an empirical investigation in order to probe more precisely the relationships among student mindware, cognitive reflection skills, and performance on challenging physics tasks.

3.1 Research Methodology

In our analyses, the Magnet question is considered challenging because it elicits persistent intuitive responses, even after instructional interventions. The screening question, on the other hand, requires a straightforward application of the same reasoning. As such, student performance on the screening question is used to gauge whether or not a student possesses necessary mindware. Finally, performance on the cognitive reflection test is used to evaluate students' cognitive reflection skills. Each correct answer to a CRT question is assigned 1 point. Students who receive a score of 2 or 3 are considered to have a relatively strong tendency toward cognitive reflection. Those who receive a score of 0 or 1 appear to exhibit tendencies toward cognitive miserliness. The following set of specific research questions was proposed. Logistic regression models were generated to probe relationships among relevant variables.

- Question 1: Is performance on the screening question linked to cognitive reflection skills?
- Question 2: Is performance on the target question before intervention linked to performance on the screening question or cognitive reflection skills?
- Question 3: Which of the following variables predict student performance on the target task after intervention: (1) student CRT score, (2) performances on the screening question, (3) performance on the target question before intervention, or (4) successful performance on the target question at any stage?
- Question 4: Does a link exists between cognitive reflection skills and the tendency to shift between correct and incorrect responses to the target question?

Due to a limited space, we briefly describe the key ideas of logistic regression relevant to our analyses. We encourage interested readers to consult references (Gette and Kryjevskaia 2019; Zwolak et al. 2018) for more detailed discussions of this statistical technique and its applicability to the PER. Logistic regression is used when the outcome variable is binary; the predictive variables could be of any type. The logistic regression model with k predictive variables has the form

$$\ln(\text{odds}) = \beta_0 + \beta_1 x_1 + \cdots + \beta_k x_k. \tag{1}$$

The odds are defined as

$$\text{odds} = \frac{\text{probability of event occurring}}{\text{probability of event not occurring}} = \frac{p}{1-p}. \tag{2}$$

Therefore, the probability for the model takes the form

$$p = \frac{1}{1 + e^{-(\beta_0 + \beta_1 x_1 + \cdots + \beta_k x_k.)}}. \tag{3}$$

The odds ratio, given by $\exp(\beta_k)$, is often employed to estimate the effect size in the context of logistic regression. It is used as an indicator of the change in odds resulting from a unit change in the predictor.

We also report the area under the receiver operating characteristic curve (ROC) and corresponding Cohen's d more commonly used in PER.

3.2   Results

3.2.1   Research Question 1

A logistic regression model was generated in which a CRT score was used as a predictor of success on the screening question. As shown in Table 2, there does not appear to be a statistically significant relationship between the two variables ($\chi^2(1) = 3.12, p = 0.08$). This result is consistent with the DPToR: Since the screening question does not elicit strong intuitively appealing responses, students with different levels of cognitive reflection skills should be equally likely to answer this question correctly. This finding is also consistent with results from our prior study conducted in the context of Newton's third law (Gette and Kryjevskaia 2019).

3.2.2   Research Question 2

Results of a logistic regression model suggest that students who answer the screening question correctly are more likely to answer the target question correctly as well. However, a CRT score does not appear to be a predictor of success on the target task. The model with performance on the screening question as the only predictor ($\chi^2 = 7.8, p < 0.01$) suggests that the odds for success on the target question are nearly 10 times higher for those students who answer the screening question correctly. The area

Table 2 Logistic regression models

Research question	Dependent variable	Predictors	Coefficients	Sig.	Exp(β)
Question 1	Performance on screening	Intercept	0.08	0.88	1.08
		CRT score	0.43	0.08	1.53
Question 2	Performance on target pre-intervention	Intercept	−2.43	0.03	0.1
		Performance on screening	2.55	0.02	12.8
		CRT score	−0.43	0.14	0.65
	Performance on target pre-intervention	Intercept	−3.05	<0.01	0.05
		Performance on screening	2.27	0.03	9.6
Question 3	Performance on target post-intervention	Intercept	−1.33	0.03	0.27
		Performance on target pre-intervention	1.73	0.03	5.6
		CRT score	0.82	<0.01	2.3
Question 4	Performance on target post-intervention (subset of students)[a]	Intercept	−0.74	0.23	0.48
		CRT score	0.86	<0.01	2.36

[a]Students who answered the target question correctly at least once (at any point during instruction)

under the ROC curve for this model is 0.65, which is equivalent to Cohen's $d = 0.5$ (Cohen 1988; Dunlap 1999). As such, the presence of mindware has a medium effect on student performance on the target question prior to instructional intervention.

3.2.3 Research Question 3

A hierarchical approach was applied in order to systematically eliminate predictors that do not improve the accuracy of the model. Results suggest that cognitive reflection skills and pre-intervention performance on the target question are the only variables linked to post-intervention performance ($\chi^2(1) = 4.00, p = 0.045; \chi^2(2) = 14.41, p < 0.01$). The two probability curves, shown in Fig. 5, reveal positive relationships between the probability of success on the post-intervention target task and CRT score. However, those students who reasoned correctly at the pre-intervention stage (solid curve) have a higher average probability of success on the assessment task compared to the students who did not (dashed curve). At the same time, students who failed to answer the magnet question correctly before the interventions and scored high on the CRT have nearly identical probabilities of success compared to students who gave correct responses before the interventions but scored low on the CRT. The Cohen's d for this model is $d = 0.8$, which suggests a large combined effect size for

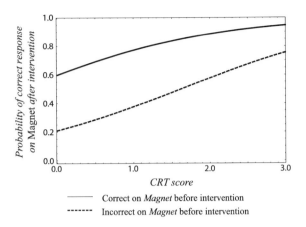

Fig. 5 Probability of correct response on target question after intervention versus CRT score

the two predictive variables. These results agree with the theoretical assertion that both mindware and cognitive reflection skills are critical for successful reasoning.

3.2.4   Research Question 4

Data from the students who gave the correct answer to the Magnet question at least once (at any stage of intervention) were included in this analysis ($N = 56$). As discussed in Sect. 2, ~30% of these students provided an intuition-based response on the target question after intervention. The average CRT score of these students is significantly lower compared to that of the students who gave a correct response on the post-test, $<CRT_{incorrect}> = 1.35$, $<CRT_{correct}> = 2.26$ (Mann–Whitney $U = 175$, two-sided $p < 0.01$). A logistic regression model in Table 2 further supports the claim that cognitive reflection skills are linked to the stability of student reasoning approaches ($\chi^2(1) = 8.85, p < 0.01$). Figure 6 illustrates that students with a low CRT score are much more likely to revert to intuitive reasoning rather continue to reason correctly. Students with stronger cognitive reflection skills, on the other hand, appear

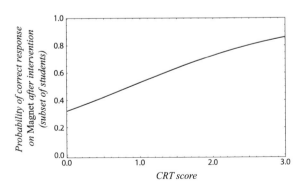

Fig. 6 Probability of correct response on target question after intervention versus CRT score (subset of students was included)

to have ~70% chance (or higher) to apply correct reasoning on the test. The area under the ROC for this mode, AUC = 0.74, suggests a large effect of cognitive reflection skills on the stability in reasoning (Cohen's $d = 0.9$). This trend is consistent with the results illustrated in Fig. 5 as well.

3.3 Discussions

Our analysis revealed that the presence of mindware does not depend on the level of cognitive reflection skills. This finding is not surprising because cognitive refection skills are not critical for learning and applying physics principles in situations that do not elicit strong intuitively appealing responses. As discussed in Sect. 2, research in cognitive psychology suggests that the CRT is not a measure of intelligence, cognitive ability, or mathematical skills (Toplak et al. 2011b; Stanovich 2009; Pennycook et al. 2016; Campitelli and Gerrans 2014).

The results from research Questions 2 and 3 may seem to be contradictory. A CRT score was determined to be a predictor of performance on the target question after intervention, but not a predictor of performance on a nearly identical target question administered before intervention. These results, however, are consistent with the DPToR. When a novice encounters the target question for the first time, the analytic system may not be placed on alert. Most students, regardless of their CRT score, appear to be confident in their incorrect, but intuitively appealing, responses. In other words, they are in a state of "cognitive ease (Kahneman 2011)." They are not tempted to reconsider their answers or look for alternatives. During the intervention, the exposure to the correct argument that challenges their strongly-held erroneous ideas may help students recognize circumstances under which intuitively appealing responses may be incorrect. Similar circumstances encountered afterward may immediately induce a state of "cognitive strain" (or alert), which is necessary for the productive engagement of the analytic process (Kahneman 2011). Students with stronger cognitive reflection skills may be more likely to recognize and act upon the state of cognitive strain. Students with weaker cognitive reflection skills may still struggle to recognize warning signs in their reasoning approaches. We argue that the instructional intervention designed and implemented as part of this investigation does appear to be productive in helping some students recognize and override instances of intuitive thoughts. However, benefits appear to be higher for those students who already possess stronger cognitive reflection skills. In addition, it is still an open question whether or not these students would be able to recognize warning signs in contexts that require farther transfer.

We were able to establish a link between cognitive reflection skills and students' shifts between formal and intuitive arguments. Again, this result is consistent with the DPToR and could be explained by the differences in student abilities to recognize and act upon red flags in their reasoning. Students with stronger cognitive reflection skills are more reliable in detecting reasoning biases. As such, they are less likely to revert back to incorrect and intuitively appealing responses.

4   Conclusions

This study is motivated by an emerging body of research suggesting that some students reason incorrectly even though they possess the relevant knowledge and skills (mindware). In order to probe the nature of inconsistent student responses even in the presence of necessary mindware, we conducted an empirical investigation guided by the dual process theories and the accompanying ideas of mindware and cognitive reflection. The theories assert that the presence of mindware is necessary but not sufficient for productive reasoning. When faced with a situation that elicits an intuitively appealing, but incorrect, response, reasoners must first resist the urge to accept such response as correct (i.e., engage in cognitive reflection). Only then they have an opportunity to evaluate their response by utilizing the necessary mindware. If a reasoner tends to rely on his or her intuition, the presence of mindware becomes less relevant.

We hypothesized that if mindware and strong cognitive reflection skills are necessary for productive reasoning, then we should be able to detect relationships among (1) student performance on physics tasks that elicit incorrect intuitive responses, (2) the presence of mindware, and (3) cognitive reflection skills. In addition, we probed whether or not an instructional intervention that takes into account student tendencies toward cognitive reflection can alter these relationships; and, if so, how. We applied the screening–target methodology in the context of forces and Newton's second law. Successful performance on the screening question implied the presence of necessary mindware. Successful performance on the target question indicated the ability to utilize this mindware in a situation that elicits intuitively appealing, but incorrect, ideas. Performance on the cognitive reflection test was used as a measure of student cognitive reflection skills.

Our analysis revealed that CRT score does not predict student performance on the screening question. This result is consistent with the DPToR: Because the screening question does not elicit strong intuitive responses, students with different levels of cognitive reflection skills are equally likely to answer the question correctly.

We found that performance on the target question before intervention is linked to performance on the screening question, but not to CRT score. This result is also consistent with the DPToR. Since the screening and target questions require the same reasoning steps, the correlation in performance is expected. (The effect size of this relationship is medium.) Even though the intuitive response to the target question is inconsistent with the formal reasoning, it was quickly perceived by many students as similar (if not identical) to correct solutions to other tasks practiced in the course. As such, many students, regardless of their CRT scores, perceived no need to give this intuitive response any further thought.

It was established that, after intervention, performance on the target question is linked to performance on the screening question and to CRT score. (The effect size of this relationship is large.) It appears that instructional interventions helped students with high CRT scores recognize warning signs in their reasoning, which, in turn, led to a more productive engagement of the analytic process. Students with low CRT scores, on the other hand, were much less successful. While a fraction of these students did give correct responses to the target question on the test, some students shifted their reasoning from correct before intervention to intuition-based after intervention. In fact, our analysis suggests that CRT score has a large effect on the tendency to shift between correct and incorrect responses.

The results of this study highlight the critical role of mindware and cognitive reflection skills in student reasoning and performance in physics. While the set of instructional interventions appears to be effective at improving performance on the target question, students with stronger cognitive reflection skills seem to benefit most. It remains an open question how to design instructional interventions successful at promoting spontaneous cognitive reflection in all students and how to facilitate farther transfer to other contexts as well.

Acknowledgements This material is based upon work supported by the National Science Foundation under Grants No. DUE-1431940, DUE-1431541, DUE-1431857, DUE-1432052, DUE-1432765, DUE-1821390, DUE-1821123, DUE-1821400, DUE-1821511, DUE-1821561. The author also wishes to acknowledge contributions from and valuable discussions with MacKenzie R. Stetzer, Beth A. Lindsey, Paula R. L. Heron, and Andrew Boudreaux.

References

Campitelli G, Gerrans P (2014) Does the cognitive reflection test measure cognitive reflection? A mathematical modeling approach. Mem Cognit 42:434

Cohen J (1988) Statistical power analysis for the behavioral sciences. Routledge, New York

diSessa AA (1993) Toward an epistemology of physics. Cogn Instr 10:105

Dunlap WP (1999) A program to compute McGraw and Wongs common language effect size indicator. Behav Res Meth Instrum Comput 31:706

Evans JSBT (2006) The heuristic-analytic theory of reasoning: extension and evaluation. Psychon Bull Rev 13:378

Frederick S (2005) Cognitive reflection and decision making. J Econ Perspect 19:25

Gette C, Kryjevskaia M (2019) Establishing a relationship between student cognitive reflection skills and performance on physics questions that elicit strong intuitive responses. Phys Rev Phys Educ Res 15:0110118

Goos M, Galbraith P, Renshaw P (2002) Educ Stud Math 49:193–223

Johnson-Laird PN (2006) How we reason. Oxford University Press Inc, New York

Kahneman D (2011) Thinking, fast and slow. Farrar, Strauss, & Giroux, New York

Kryjevskaia M (2019) Examining the relationships among intuition, reasoning, and conceptual understanding. In: Pietrocola M (ed) Physics upgrading physics education to meet the needs of society. Springer, Cham, pp 181–188

Kryjevskaia M, Stetzer MR, Heron PRL (2012) Student understanding of wave behavior at a boundary: the relationships among wavelength, propagation speed, and frequency. Am J Phys 80(4):339–347

Kryjevskaia M, Stetzer MR, Grosz N (2014) Answer first: applying the heuristic-analytic theory of reasoning to examine student intuitive thinking in the context of physics. Phys Rev Spec Top – Phys Educ Res 10, 20109

Kryjevskaia M, Stetzer MR, Le TK (2015) Failure to engage: examining the impact of metacognitive interventions on persistent intuitive reasoning approaches. In: Proceedings of 2014 physics education research conference. AAPT, College Park, pp 143–146

Lindsey BA, Nagel MA, Savani BN (2018) Leveraging physics understanding of energy to overcome unproductive intuitions in chemistry. Phys Rev PER (submitted for publication)

Lising L, Elby A (2005) The impact of epistemology on learning: a case study from introductory physics. Am J Phys 73:372

Nickerson RS (1998) Confirmation bias: a ubiquitous phenomenon in many guises. Rev Gen Psychol 2:175

Pennycook G, Cheyne JA, Koehler DJ, Fugelsang JA (2016) Is the cognitive reflection test a measure of both reflection and intuition? Behav Res Methods 48:341

Sabella M, Cochran G (2004) Evidence of intuitive and formal knowledge in student responses: examples from the context of dynamics. In: Marx JD, Cummings K, Franklin SV (eds) Proceedings of 2003 physics education research conference. AIP conference proceedings, vol 720. Madison, WI, pp 89–92

Simon HA (1992) What is an explanation of behavior? Psychol Sci 3:150

Singh C (2002) When physical intuition fails. Am J Phys 70:1103

Stanovich KE (2009) What intelligence tests miss: the psychology of rational thought. Yale University Press, Yale

Thompson VA, Prowse Turner JA, Pennycook G (2011) Intuition, reason, and metacognition. Cogn Psychol 63:107

Thompson VA, Evans JSBT, Campbell JID (2013) Matching bias on the selection task: it's fast and feels good. Think Reason 19:431

Toplak ME, West RF, Stanovich KE (2011a) The Cognitive Reflection Test as a predictor of performance on heuristics-and-biases tasks. Mem Cognit 39:1275

Toplak ME, West RF, Stanovich KE (2011b) The Cognitive Reflection Test as a predictor of performance on heuristics-and-biases tasks. Think Reason 20(2):147–168

Zwolak JP, Zwolak M, Brewe E (2018) Educational commitment and social networking: the power of informal networks. Phys Rev Phys Educ Res 14:010131

Design Tools as a Way to Explicitly Connect Research Insights with Design Decision for Teaching Learning Sequences

Jenaro Guisasola, Kristina Zuza, Jaume Ametller, and Paulo Sarriugarte

Abstract This chapter discusses the concept of design tools in the context of the design of teaching learning sequences in physics education. Design tools are compared to humble theories (Cobb et al in Educ Res 32:9–13, 2003) because these theories are connected to specific, content-based, learning processes and are to the activity of design. We find this comparison useful and try to develop it further in this paper. We will present three design tools we have used in our research showing how they are used. Each of them is connected to the other ones, and we will use examples from a TLS on forces to show how they fit together in the process of design. Afterwards, we will present another design tool, communicative approaches, as an example of a design tool that we chose not to use in the design of the TLS on forces we have used as an example. In the final section, we will discuss the role of the design tools in the articulation of a research programme on TLS design.

1 Introduction: Design Tools for Designing Teaching Learning Sequences

The concept of design tools to design teaching learning sequences has been defined as "concepts which draw upon theoretical perspectives on teaching and learning, and the products of empirical research on teaching and learning, to inform decisions about the design of teaching" (Ametller et al. 2007). This definition, rooted in a line of research that takes engineering as a metaphor in science education (Hjalmarson and Lesh 2008), makes design tools a way of capturing the aim of explicitly base the design of teaching in theoretical insights and empirical results which has been

J. Guisasola (✉) · K. Zuza · P. Sarriugarte
Donostia Physics Education Research Group (DoPER), Department of Applied Physics,,
University of the Basque Country (UPV-EHU), Leioa, Spain
e-mail: jenaro.guisasola@ehu.es

J. Ametller
Department of Specific Didactics, University of Girona, Girona, Spain

J. Guisasola and K. Zuza (eds.), *Research and Innovation in Physics Education: Two Sides of the Same Coin*, Challenges in Physics Education,
https://doi.org/10.1007/978-3-030-51182-1_9

defended by several research groups (Ruthven et al. 2009). This aim is also at the core of our work on teaching learning sequences (Guisasola et al. 2017) where our aim is to design teaching learning sequences (TLS) showing explicitly how theoretical and empirical research insights are used to guide the design of the TLS so that the products and their implementation can be the object of a research programme on the design of TLS to foster the learning of particular physics (science) topics. In the pursue of that aim, design tools appear as a way to operationalise the bridge between existing knowledge on or related to science education and specific elements of TLS. Of course, it is clear from the beginning that this won't be a straightforward endeavour for design tools aim at bridging quite disparate things but this is precisely why they are so central in our design process because they make visible a bridge that seems easier to cross implicitly. Hence, with all their shortcomings, design tools open up to scrutiny, analysis, reflection and systematic improvement the processes that lie at the heart of the design of TLS in science education.

Design tools are also predicated in the assumption that the TLS design process can be "standardised" -up to a point-. We do not intend to make an exhaustive list of which design tools can be defined or want to suggest that all the design tools have to be used in the design of any TLS. We believe, however, that properly defined design tools should be available for designers to choose from according to the designer's aims. Design tools are tools that can be used generically in the design process to produce specific, content bound, elements of the TLS. "The word tool is used to underline the fact that theoretical insights are brought to bear on the design process, and real work is then carried out" (Ametller et al. 2007). Therefore, tools must be defined in a way that can lead to a certain degree of operationalisation of the theoretical and empirical insights designers chose as relevant to the teaching to be designed in order to generate, from these ideas, specific elements of the teaching or TLS being designed. Even when this is possible one should not expect anything resembling an algorithm, a degree of professional knowledge must be exercised in the process. A professional knowledge that must be informed by research and relevant theory but which will involve implicit knowledge as well. The virtue of the design tools on making that process explicit and on framing the scope of application of the professional knowledge and criteria so that designers can more easily document, evaluate and discuss the coming about of the constitutive elements of the design.

The difficulty of finding "design tools" has to do with the fact that the diverse nature of research inputs and the diversity of aspects and elements of designed teaching imply a variety of connections. When considering design experiments (Cobb et al. 2003), the design of the teaching can encompass a wide spectrum covering topic-specific classroom activities and generic pedagogical orientations on, for instance, the organisation of group work and even sociologically oriented indication for *systemic conditions*. In our case, we restrict our scope to the design of the TLS, hence focusing on class/groups and trying to tailor TLS to organisational and curricular conditions rather than suggesting changes at that level. Even then, research inputs range from psychological theories of learning to specific questions to guide a part of the learning progression of a specific topic. We will have to deal with content-specific and content-independent inputs, with fine grain and large grain size elements, correspondingly

(Leach and Scott 2008). It seems reasonable that the less topic-specific the research the more latitude it will afford the designers. Nevertheless, such latitude will not be unbounded or infinite and it will have to be in concordance with other design choices meaning that different design tools, or the use of the same design tool with different inputs, might constructively interact to better define the design decisions on relation to the research inputs.

In the literature (Ametller et al. 2007; Ruthven et al. 2009), design tools are compared to humble theories (Cobb et al. 2003) because these theories are connected to specific, content-based, learning processes and are to the activity of design. We find this comparison useful and try to develop it further in this chapter.

Once we have discussed the concept of design tool, we need to consider which specific design tools can be defined. We propose that this question might be approached by looking at which elements of the TLS one can intend to base of research, selecting the type of research that can be used for a particular element of design and then propose a way to based the former on the latter. How much of the design can be based on research depends on the level at which we consider the TLS. At a large grain size, we can base our design on psychological theories but as we move to the more concrete levels of design we increasingly rely on humble theories and results from empirical research. At this level, we might be facing a fairly large number of decisions that we might connect to research insights but we must also acknowledge that the evidence backing some areas of physics education research are too weak, or context-dependent, to be able to provide strong guidance, and hence, they can only be taken as inputs for the calls we will have to make during the design. It is clear that no design tools can be defined here. Hence, we should start by sticking to elements or aspects for which there is enough information and that we can argue to be important enough to warrant research informed decisions.

Starting at a large grain size, taking the most general decisions about the design of our TLS, we will start by connecting our design to the more general theories we use in the process. The decisions we take at this point will also guide some of the decisions at more concrete levels of design. This connection will be mirrored in the connection we will see among design tools operating at different grain size levels as well. In the following sections, we will present three design tools we have used in our research showing how they are used. Each of them is connected to the other ones and we will use examples from a TLS on forces to show how they fit together in the process of design. Afterwards, we will present another design tool, communicative approaches, as an example of a design tool that we chose not to use in the design of the TLS on forces we have used as an example. In the final section, we will discuss the role of the design tools in the articulation of a research programme on TLS design.

2 Three Design Tools Used in the Design of a TLS on Newton's Laws

In this section, we will present three design tools. We will define each tool and show an example of its use. The examples of application shown for the three design tools correspond to a TLS on dynamics (Guisasola et al. 2019) which allows us to present the design tools going from the larger grain size to the more specific showing how decisions on one tool impact on the next one.

2.1 Design Tool 1: Ontological—Epistemological Analysis

The ontological—epistemological analysis tool that might be misunderstood for it seems to connect not research on science (physics) education with design issues but to connect physics knowledge with the design of its teaching/TLS. What the onto-epistemological analysis does is to provide a way of defining learning objectives (conceptual learning objectives and elements of Nature of Science (NOS) which includes ontological characteristics, of learning objectives) based on an onto-epistemological analysis of the learning aims. This is both connected to research in the learning of specific contents and to the epistemology of science as one of the theories that constitute the basis of science education as a field of study. We included the onto-epistemological analysis in the area of research on History and Philosophy of Science (HPS) that provides meta-perspectives on science and includes not only HPS but also other disciplines such as technology, mathematics or sociology (Erduran 2020).

The research on specific contents shows that a lack of a proper epistemological analysis might be connected to imprecise or incorrect conceptual (or NOS related) approaches, which might be present even when the overall aim is consistent with the curriculum. Hence, the (onto) epistemological analysis provides a research-based justification for the definition of the learning aims:

> The learning aims are justified not only by the curriculum or the teaching tradition but also by the arguments that the epistemology of physics justify as essential to build a scientific model (Guisasola et al. 2019).

In summary, the onto-epistemological analysis as a design tool provides a way of clearly identifying key science ideas which construct the conceptual content to be taught and to define from it the leaning aims at a specific level so that it starts to describe the fundamental structure of the hypothesised learning pathways.

As an example of the use of the onto-epistemological analysis design tool, we will show how we defined some of the research aims of a TLS on Newton's laws (Guisasola et al. 2019). The first step is to define the essential elements of the topic for its teaching at the secondary school level. To do so, we look at the historical development of the ideas of forces (in classical physics) and how they have been described from the epistemology perspective by other authors (Coelho 2010; Eisenbud 1958; Ellis 1962). This leads to the identification of key ideas for this topic:

K.1. The concept of fore is a measurement of the interaction between two bodies A and B. Therefore, an isolated body cannot experience a force.

K.2. The force exerted by a body A on body B has exactly the same magnitude that the exerted by B on A and the opposite direction than the force exerted by B on A. They do so simultaneously (Newton's third law).

K.3. Forces exerted on a body produce acceleration or a deformation.

K.4. Newton's second law relates the force exerted on a body and the acceleration it acquires through the magnitude of inertial mass.

K.5. The dynamic of circular motion requires the vector nature of the force and its acceleration.

Taken together, these key ideas can be used to define the learning aims of a TLS on classical dynamics for secondary school students—according to the national curriculum for which the TLS was designed. These aims are not a one-to-one correspondence with the key ideas but an elaboration that requires professional knowledge on how to define learning aims. In our example, the first three key ideas were translated into the following learning objectives:

LO 1.1. To understand forces as interactions between two bodies

LO 1.2. To understand that forces between two bodies have the same magnitude and opposite direction (Newton's third law). Moreover, to recognise that no force precedes the other in time.

LO 1.3. To recognise that forces have a vector nature and to be able to apply it when drawing force diagrams.

From this example, we clearly see that this tool is intended to move from epistemological constituents to conceptual learning aims. The development of these aims requires professional knowledge regarding curriculum contents and how it relates to the epistemological analysis as well as on defining good learning aims, aims that are actionable.

The onto-epistemological analysis might be found elsewhere if it exists or has to be carried out by member of the design team. It is also the case that divergencies between (onto) epistemological analysis and implemented proposals provide insights on the reasons for some of the difficulties encountered.

2.2 Design Tool 2: Learning Demand

While the (onto) epistemological analysis is based on the conceptual basis of physics, the theory underpinning learning demands is educational psychology theory; hence, both tools taken together are underpinning the design of the TLS on fundamental theoretical foundations of physics education. The learning demand tool is based on socio constructivism (Leach and Scott 2008) in particular the concepts of previous or alternative conceptions, zone of proximal development (ZPD) and conceptual change.

According to these constructivist concepts, students' learning is constructed starting by ideas he or she already has and then will undergo a process of conceptual change. To foster this process, science teaching must be aimed at designing and implementing activities which place a demand in the students within their ZPD. This ZPD is defined as that which students can do on their own and that which they can do with the assistance of an expert. Hence, the ZPD can be seen as the distance between what students know and what they need to achieve. How big a gap that is not a direct measure of how difficult is the learning aim but it is clearly connected. The learning demand measures the ontological and epistemological distances between what students know (their previous ideas) and what we have as a learning goal, in the specific TLS we are designing.

Therefore, the learning demand design tool compares the learning aims—which we have defined through the use of the onto-epistemological analysis design tool—and the students ideas about the topics addressed in those aims. The students' ideas can usually be found in the extensive research literature of previous conceptions or, when this is not available for a particular topic, the designer team needs to carry out a research to describe them. Once both inputs have been identified the designers need to establish the ontological and epistemological differences between the starting point and the final aim of the intended learning process. This gap is the definition of the learning demand (as a concept). At this point, designers decide the "measure" of that gap as an expression of how difficult it is for students to bridge it. This decision is informed by the professional knowledge of designers and by the empirical results on students' learning of those topics when they exist.

The measure of the leaning demand is used then to make decisions on where to put the teaching efforts. The larger the demand the longer it might require as well as taking into account that the ZPD must require more steps. The type of the demand helps designers decide the focus of the activities that will be included in the TLS. We concrete the level of the learning demand regarding the quantity and/or deepness of the activities of TLS. Difficulties related to small learning demands are addressed by some simple activities. When the required learning demand is high, the activities must be designed upon active learning methodologies and the proposed tasks must be complex enough to help students to overcome the specific difficulty. The socio-constructivist theory guides decisions on types of activities and sequencing connected to the measurement of the learning demands but does not determine the specific activities that need to be included in the TLS, and this is a finer grain size level that we will address later on.

Coming back to the example of the TLS on Newton's laws, we took the learning aims defined with the onto-epistemological design tool and compared them with the students' previous ideas on those topics found in the literature. We judged that the most relevant ones are (Ellis 1962; Savall-Alemany et al. 2019; Guisasola et al. 2008):

PI 1 Secondary school students have difficulties when working with vectors
PI 2 Larger bodies exert larger forces.
PI 3 Only forces that can be felt have an effect. The bodies exert forces
PI 4 Friction forces lead moving bodies to rest.

Table 1 Learning demands associated with the learning aims defined with the onto-epistemological design tool

Defined learning objectives	Previous ideas found in the literature	Type and measure of learning demand
LO1.1. To understand forces as interactions between two bodies LO1.2. To understand that forces between two bodies have the same magnitude and opposite direction (Newton's third law). Moreover, to recognise that no force precedes the other in time LO1.3. To recognise that forces have a vector nature and to be able to apply it when drawing force diagrams	• Secondary school students have difficulties when working with vectors • Larger bodies exert larger forces. The bodies exert forces • Only forces that can be felt have an effect • Friction forces lead moving bodies to rest	Ontological—medium Ontological and Epistemological—medium Ontological—small Ontological and epistemological—large

When comparing these previous ideas with the learning aims, we had defined we established both the type (ontological or epistemological) and the measure (small, medium or large) of the learning demand. In Table 1, we have brought together this information associating the learning demand to the previous ideas.

2.3 Design Tool 3: Driving Problems

So far we have seen how two design tools that have been used to operationalise insights from epistemology of physics and educational psychology into the conceptual leaning objectives and he focuses on the learning activities, respectively.

The third source of insights into physics education is pedagogy. Here we will look for insights on how to orientate the activities we have to design in terms of structuring the TLS proposal and deciding on teaching strategies. This implies moving into a finer grain size, and in this case, it also implied moving into a knowledge that is highly contextual and akin to humble theories at best. Consequently, the distance between the research information we use to base our design and the specific design choices grows smaller. Nevertheless, it is possible to define design tools at this level as well. The third design tool, we present in this chapter, the driving problems, is an example of such a tool. Driving problems refer to the definition of questions or problems that structure the TLS activities and sequence the intended learning progressions. Driving problems are related to a number of teaching-learning strategies centred on the resolution of meaningful problems, such as PBL to give one of the better-known proposals. These approaches are being increasingly used in science education because research on their use shows comparatively good learning results (Savall-Alemany et al. 2019). Furthermore, by articulating the learning of physics we relate to both the nature of science and the active character of learning in socio-constructivist approaches which

Table 2 Learning indicators and corresponding driving questions

Learning Objectives	Driving Questions
LI 1.1. To understand forces as interactions between two bodies LI 1.2. To understand that forces between two bodies have the same magnitude and opposite direction (Newton's third law). Moreover, to recognise that no force precedes the other in time LI 1.3 To recognise that forces have a vector nature and to be able to apply it when drawing force diagrams	DQ 1. What is a force? DQ2. How can we represent it and measure it?

makes problem-based strategies consistent with the theoretical referents that we have used so far to underpin the design of the TLS.

Driving problems as a design tool bring together what we know about the characteristics of good structuring questions (Guisasola et al. 2008) to make a proposal of driving questions that operationalise into a sequence of activities to guide a learning progression to achieve the learning indicators taking into account the logical structure revealed in the onto-epistemological analysis, the type and measure of the identified learning difficulties and the research results on teaching the proposed topic. In our example of the TLS on Newton's laws, this has led to the driving questions presented in Table 2.

3   The Role of Design Tools in a Research Programme on Teaching Learning Sequences

The previous sections have shown how we have used three design tools in the process of designing a TLS on Newton's laws for secondary school physics. In this final section, we will briefly discuss other examples of design tools present in the literature. We will address their place in the development of a research program on teaching learning sequences for physics education, both as design elements and as research outcomes.

The three design tools we have used in the example presented in the previous section are not the only design tools that can be found in the literature. Some of them address different aspects of the design of teaching while others have many points in common with the ones we use. An example of the former is the communicative approach tool (Mortimer and Scott 2003) which guides decisions on classroom discourse based on sociocultural learning theory. The tool allows designers to suggest how the discourse should change during the designed teaching-learning process to match the stages of learning according to the sociocultural theory (Scott and Ametller 2007). While this design tool is addressing the staging of teaching, it can also be included in the TLS (Scott et al. 2006) to provide that information to teachers. We have not included this detailed information on the intended discourse in our TLS but the commonalities of the theoretical underpinning of this tool and our tools mean that we could include it in our design process in the future.

Another example of design tools is the knowledge distance tool proposed by Thibergien and colleagues (Buty et al. 2004), which makes explicit the difference between the contents of the curriculum students need to learn and their existing knowledge. This tool is clearly similar to the learning demand tool. Even though the focus of their analysis is different—models for the former and social discourses in the latter—and they have different approaches to taking curriculum contents into consideration, their common theoretical background makes it easy to imagine bringing them together into a single design tool (Ruthven et al. 2009).

These two examples raise the question of how many design tools could be defined and how many should be considered for the design of any given TLS. Answering in depth of these questions goes beyond the scope of this chapter, but we can offer a tentative answer in line with our general aims as we have presented them in Sect. 1. A brief overview of the literature on the design of physics teaching, in particular TLS, shows that most design tools are connected to the epistemology of physics and socio-constructivist perspectives of learning. These commonalities bring about design tools that are often addressing a similar design point and hence could converge, or address different elements -within the same general framework—and hence could be complementary. As we have already said, it is our view that it would be desirable to have a widely shared approach to the design of TLS to articulate a research program in this area. With this aim in mind the convergence of similar design tools would be desirable, not just to simplify the discussion in the field, but because the process of redefining convergent tools would likely help clarifying and making more explicit the connection between the design tools and the research underpinning them. Complementary tools might address elements that are more or less relevant depending on the context of the application and hence should be used or not according to the professional criteria of the designers. On the other hand, if enough evidence is gathered through research backing the effectiveness of a particular design tool and the design decisions derived from it, designers should strive to include them in their design of TLS.

The three design tools that we have used in our example have been originally developed during the past two decades by two research groups, to which authors of this chapter have been part, working on the design of teaching. This particular choice of tools has forced us to reconsider how we had used them so far, help us clarified their definition and use, and provided us with a set of design tools that have allowed us to improve the design process and make it more explicit. These choices represent the convergence of lines of research in this area that we want to encourage in order to build a shared research programme on the design of teaching learning sequences.

Design tools are, hence, a key element of our proposal for the research-based design of TLS in physics education. We are convinced that they are essential to bridge the large and the fine grain size aspects of the TLS design and to provide a common language required to have a shared research programme.

Design tools should also be the focus of particular attention in the evaluation of the designed TLS so that designers/researchers would have enough information on how they impact their effectiveness. Design tools should be, therefore, part of the research results of the TLS research programme alongside humble theories. The latter will continue to provide insights on content-based teaching and learning processes while design tools will provide insights on connecting grand theories and humble theories to guide the design of any physics TLS.

References

Ametller J, Leach J, Scott P (2007) Using perspectives on subject learning to inform the design of subject teaching: an example from science education. Curriculum J 18:479–492

Buty C, Tiberghien A, le Maréchal JF (2004) Learning hypotheses and an associated tool to design and to analyse teaching-learning sequences. Int J Sci Educ 26:579–604

Cobb P, Confrey J, diSessa A et al (2003) Design experiments in educational research. Educ Res 32:9–13

Coelho RL (2010) On the concept of force: how understanding its history can improve physics teaching. Sci Educ 19:91–113

Eisenbud L (1958) On the classical laws of motion. Am J Phys 26:144–159

Ellis BD (1962) Newton's concept of motive force. J Hist Ideas 23:273–278

Erduran S (2020) Editorial vision for science and education. Science and Education, pp 1–5

Guisasola J, Furió C, Ceberio M (2008) Science education based on developing guided research. In: Thomase MV (ed) Science education in focus. Nova Science Publishers, Hauppauge, New York, pp 55–85

Guisasola J, Zuza K, Ametller J et al (2017) Evaluating and redesigning teaching learning sequences at the introductory physics level. Phys Rev Phys Educ Res 13. Epub ahead of print 2017. https://doi.org/10.1103/physrevphyseducres.13.020139

Guisasola J, Zuza K, Sagastibeltza M (2019) Una propuesta de diseño y evaluación de secuencias de enseñanza-aprendizaje en Física: el caso de las leyes de Newton. Revista de Enseñanza de la Física 31:57–69

Hjalmarson M, Lesh R (2008) Engineering and design research: Intersections for education research and design. In: Kelly AE, Lesh RA, Baek JY (eds) Handbook of design research methods in education: Innovations in science, technology, engineering, and mathematics learning and teaching. Routledge, London, pp 96–110

Leach J, Scott P (2008) Teaching for conceptual understanding: an approach drawing on individual and sociocultural perspectives. In: Vosniadou S (ed) Interantional handbook of research on conceptual change. Routledge, London, pp 647–675

Mortimer E, Scott P (2003) Meaning making in secondary science classrooms. Open University PRess, Maidenhead

Ruthven K, Laborde C, Leach J et al (2009) Design tools in didactical research: Instrumenting the epistemological and cognitive aspects of the design of teaching sequences. Educ Res 38:329–342

Savall-Alemany F, Guisasola J, Rosa Cintas S et al (2019) Problem-based structure for a teaching-learning sequence to overcome students' difficulties when learning about atomic spectra. Phys Rev Phys Educ Res 15:20138

Scott P, Ametller J (2007) Teaching science in a meaningful way: striking a balance between "opening up" and "closing down" classroom talk. Sch Sci Rev 88:324

Scott P, Leach J, Hind A et al (2006) Designing research evidence-informed teaching interventions. In: Millar R, Leach J, Osborne J et al (eds) Improving subject teaching: lessons from research in science education. Routledge, London, pp 60–78

Results of a Design-Based-Research Study to Improve Students' Understanding of Simple Electric Circuits

Jan-Philipp Burde and Thomas Wilhelm

Abstract Most secondary school students fail to develop an adequate understanding of electric circuits as they tend to reason exclusively with current and resistance. Effective reasoning about electric circuits, however, requires a solid understanding of the concept of voltage. Against this background, a new teaching concept based on the electron gas model was developed with the goal to give students a qualitative but robust conception of voltage as a potential difference that causes the electric current. Using an air pressure analogy, the teaching concept aims to provide students with intuitive explanations that have their origins in the students' everyday experiences, e.g. with bicycle tires or air mattresses. Similarly to these everyday objects, where air pressure differences cause an airflow, voltage is introduced as an electric pressure difference across a resistor that causes the electric current. An empirical evaluation with 790 secondary school students shows that the new teaching concept leads to a significantly better conceptual understanding than traditional teaching approaches in Germany. Furthermore, 12 of the 14 participating teachers state that they plan to teach according to the new concept in future as they consider it to be a significant improvement.

J.-P. Burde (✉)
Physics Education Research Group, University of Tübingen, Auf der Morgenstelle 14, Tübingen 72076, Germany
e-mail: Jan-Philipp.Burde@uni-tuebingen.de

T. Wilhelm
Department for Physics Education Research, Goethe University Frankfurt, Max-von-Laue-Str. 1, 60438 Frankfurt am Main, Germany
e-mail: wilhelm@physik.uni-frankfurt.de

J. Guisasola and K. Zuza (eds.), *Research and Innovation in Physics Education: Two Sides of the Same Coin*, Challenges in Physics Education,
https://doi.org/10.1007/978-3-030-51182-1_10

1 Motivation

Having taught the topic of simple electric circuits, most teachers will find that despite all their efforts, many students fail to develop an adequate understanding of voltage and electric circuits in general. Not realising the important role of voltage, students tend to reason exclusively with current and resistance when dealing with electric circuits (Cohen et al. 1983). As a result, they often have a series of alternative conceptions about electric circuits and generally struggle to understand how circuits work (Duit et al. 1985; Wilhelm and Hopf 2018). In particular, students often think of voltage as a property or a component of the electric current rather than an independent physical quantity that refers to a difference in electric potential (Rhöneck 1986). As students struggle to distinguish between the electric current and voltage, the former often dominates their understanding of electric circuits. As a consequence, these students see no need to conceptualise voltage as an independent physical quantity and hence fail to realise the important relation of cause and effect between voltage and current.

1.1 Background

Although the reasons for these learning difficulties are complex and numerous, three main problems can be identified based on prior research in science education: Firstly, for historical but not educational reasons, the concept of the electric current dominates teaching at the expense of potential and potential difference. For that reason, Cohen, Eylon & Ganiel point out that *"we need a curriculum that introduces the concept of potential difference first and [...] clearly spells out the relation of cause and effect between pd* [potential difference] *and current"* (Cohen et al. 1983). Secondly, alternative explanations of voltage (e.g. $V = \Delta E/q$) make it difficult for the students to understand the mutual relationship of V, I and R in electric circuits as well as the fact that voltage represents a potential difference (Härtel 2012; Herrmann and Schmälzle 1984). Thirdly, an extensive but purely quantitative study of the formula $V = R \cdot I$ in physics lessons is highly problematic as "[...] *premature mathematization and 'exact' definitions* [often distort] *a conceptual understanding without really being able to replace it"* (Muckenfuß and Walz 1997, translation by the authors). In particular, the focus on the formula $V = R \cdot I$ can even strengthen the students' misconception that voltage must be a property of the electric current since the formula suggests that both physical quantities can only occur simultaneously (Muckenfuß and Walz 1997).

A more general problem lies in the fact that the physical processes in electric circuits are quite abstract and hard to imagine for students, because the electron movement, for example, is beyond direct perception. A way to help students understand the abstract concepts of electricity is to use models of electric circuits. While good models and analogies can indeed foster a deeper conceptual understanding, physics education research has shown that—contrary to popular belief—the use of the widespread water circuit analogy using closed water pipes can reinforce typical alternative conceptions (Schwedes et al. 1995; Schwedes and Schilling 1983). Although the analogy is undoubtedly quite powerful from a purely physical point of view, the problem with it lies in the fact that students have no experience with closed water circuits from their everyday lives. In particular, they have no experience with water pressure in water pipes and think of water as an incompressible fluid. Since water under high pressure differs neither visibly nor palpably from water under low pressure, the water circuit analogy has proven less compelling to learners than generally expected (Burde and Wilhelm 2016). In contrast, the introduction of voltage as a potential difference has proven to be comparatively effective in promoting learning in a number of studies (Waltner et al. 2009; Gleixner 1998; Schumacher and Wiesner 1997). Examples of models of electric circuits that introduce voltage as a potential difference include the "rod model" developed in Munich by Gleixner (Gleixner 1998) and air pressure analogy used in the CASTLE curriculum by Steinberg and Wainwright (Steinberg and Wainwright 1993). While the former is suitable for illustrating potential differences, the latter has proven to be particularly promising to explain the relationship between potential difference and current as the electric potential is compared to air pressure. The advantage of the air pressure analogy over the water pressure analogy is that it has proven to be a highly intuitive and yet powerful analogy as students have a variety of experiences with air pressure from their everyday lives, e.g. with air mattresses, footballs or bicycle tyres (Burde and Wilhelm 2016).

1.2 Shortcomings of the CASTLE Curriculum

Although the CASTLE curriculum with its underlying air pressure analogy undoubtedly represents a promising approach to teaching electric circuits, there are several shortcomings regarding its design and evaluation. For example, in contradiction to the considerations made above, the CASTLE curriculum introduces the electric current before potential differences. As outlined before, this traditional content structure may in fact prevent students from understanding the important role that potential differences play in electric circuits as too much emphasis is placed on the electric current and too little emphasis is placed on the cause-effect-relationship between voltage and current. In this context, Cohen, Eylon & Ganiel point out that "*first impressions are strong and may impede a later, more rigorous, study of electricity*" (Cohen et al. 1983). Another drawback of the CASTLE curriculum lies in the fact that it primarily consists of a series of hands-on experiments with specially designed capacitors. In regard to the German school system, this does not only represent a problem as such capacitors can usually not be found in sufficient numbers in German physics classes, but also because the traditional lesson time of 45 min is generally ill-suited for a teaching concept that primarily relies on hands-on experiments. Another point of criticism is that the learning effectiveness of the CASTLE curriculum has never been empirically evaluated. Although its authors claim that it leads to significantly larger achievement gains than traditional approaches to teaching electric circuits (Steinberg and Wainwright 1993), no empirical results of a study on its learning effectiveness have been published yet.

1.3 Goals and Research Questions

What has been lacking to date is a teaching concept that is based on the powerful air pressure analogy, but which introduces the concept of potential difference before the electric current and which is compatible with German school standards. What has also been lacking to date is an empirical evaluation of the learning effectiveness of such a teaching concept. In this paper, we will therefore first focus on the key ideas of the new teaching concept before presenting the results of its empirical evaluation. Here, we will particularly focus on the question whether the new teaching concept leads to a better conceptual understanding than traditional approaches to teaching electric circuits and whether the participating teachers consider the new teaching concept to be an improvement of their teaching practice.

2 The New Teaching Concept

Since the 1970s, a lot of research has been conducted on students' learning and students' alternative conceptions in introductory electricity (Duit et al. 1985; Wilhelm and Hopf 2018). However, the insights gained by numerous studies on domain-specific learning rarely had an adequate impact on teaching practice. From our perspective, this can at least partly be explained by the fact that knowing about students' alternative conceptions itself is not enough for teachers to successfully promote conceptual change in the classroom. In order to overcome typical alternative conceptions and successfully trigger conceptual change, educators need well-devised teaching resources that incorporate relevant research findings from the physics education community. While research results on students' alternative conceptions are specifically taken into account, diSessa's perspective on learning as the construction and reorganisation of previously only loosely connected elements of knowledge, called "p-prims", into a coherent mental structure forms the theoretical foundation of the teaching concept (Burde and Wilhelm 2018). These p-prims ("phenomenological primitives") represent a fragmentary and naive understanding of the physical world. They are primitive in the sense that they only constitute minimal abstractions from everyday experience (diSessa 1993). As diSessa points out, successful conceptual change can only occur if the students' prior knowledge in the form of p-prims is taken into account: "Students have a richness of conceptual resources to draw on. Attend to their ideas and help them build on the best of them" (diSessa 2008). A comprehensive description of diSessa's perspective on learning can be found in diSessa (2008).

In accordance with diSessa's perspective on learning, the new teaching concept aims to foster conceptual change by building on students' everyday physical intuitions, in this case with air pressure. In contrast to traditional approaches to teaching electric circuits, the new teaching concept does not focus on quantitative aspects, but on a qualitative understanding of voltage, current and resistance as well as their mutual relationship by constantly providing students with intuitive explanations that have their origins in the students' everyday experiences. Since the electric current generally seems to dominate students' understanding of electric circuits, a main objective of the new teaching concept is to establish potential differences as the starting point of any analysis of electric circuits. In accordance with the considerations of Cohen, Eylon & Ganiel, the objective is to make voltage and not the electric current the students' primary concept when thinking of circuits (Cohen et al. 1983).

2.1 Air Pressure Differences Cause an Airflow

In order to achieve this objective, the teaching concept builds on the students' intuitive concept of air pressure in the sense that "compressed air is under pressure, pushes against the walls and tries to expand". At the example of everyday objects such as bicycle tyres and air mattresses, students learn that air always flows from areas of high pressure to areas of low pressure. The conclusion then is that pressure differences are the cause for an airflow and that a conceptual distinction must be made between pressure and pressure difference. The explicit discussion of air pressure phenomena takes place against the background that learners often have conceptual difficulties in distinguishing between pressure and pressure difference. Towards the end of the unit on air pressure, a first concept of resistance is introduced, with students taking a piece of fabric (e.g. a scarf, collar or sleeve) and blowing air through it. By doing so, they learn that the thicker the piece of fabric is folded, the stronger the inhibition or obstruction of the airflow is (see Fig. 1). The inhibition or obstruction of the airflow by the fabric is then referred to as "resistance".

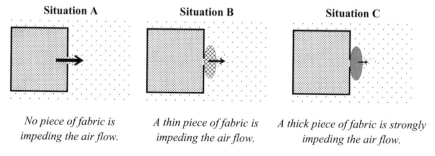

Fig. 1 Pressure differences cause an airflow with a piece of fabric impeding the airflow

2.2 Battery, Electric Potential and Voltage

In the next unit, the idea of air pressure is transferred to the electric circuit by assuming that electrons, as particles, can move freely in a conductor, where they form an "electron gas". Since the electrons are negatively charged, they are pushed apart as far as possible by repulsion, which is why they uniformly fill the space available to them in the entire conductor. Due to the mutual Coulomb repulsion of the electrons, an electric pressure dependent on the electron density results. By assuming a surplus of electrons at the negative terminal of a battery and a shortage of electrons at its positive terminal, it is then argued that there is a high electric pressure at the negative terminal and the wire connected to it and a low electric pressure at the positive terminal and the wire connected to it.

In order to visually emphasise the similarity between air pressure and electric pressure, the dot-density representation already known from the air pressure examples is initially also used for open electric circuits (see Fig. 2, left). From this point on, however, it is better to use colour coding to visualise the electric pressure instead of the dot-density representation, since colour coding the electric pressure using coloured pencils has proven to be a more practical and timely method (see Fig. 2, right). In addition, colour coding also avoids the impression that the resistor consumes the moving electrons. In contrast to the often rather unsystematic choice of colours in existing teaching concepts, the colour scheme used in the concept presented in this paper is based on a convention that students should be familiar with from their everyday lives. Similarly to temperatures that are usually represented by the two colours red (high temperatures) and blue (low temperatures), e.g. in weather reports, thermal imaging cameras and water taps, a high electric pressure is represented by red, whereas a low electric pressure is represented by blue in this teaching concept. Since no absolute values are specified for the electric pressure, electrical grounding is deliberately not covered in the teaching concept.

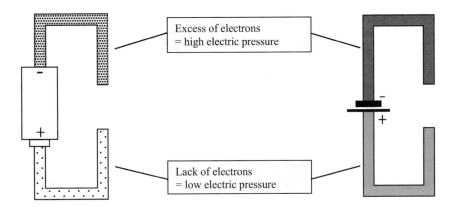

Fig. 2 Dot-density representation (left) and colour coding (right) of the electric pressure

2.3   Electric Current and Resistance

In analogy to the air pressure examples discussed before, electric pressure differences are introduced as the cause of the electric current. For this purpose, a simple circuit consisting of a battery and a light bulb is used to discuss that the bulb will light up because the electric pressure difference across it will cause an electric current through the light bulb (see Fig. 3).

Based on the concept of resistance acquired in the unit on air pressure, the students are then given a qualitative idea of resistance in electric circuits and how it affects the electric current. Here, students learn that a resistor impedes the electric current in the same way as a piece of fabric impedes an airflow. The influence of voltage on electric current as well as of electrical resistance on electric current is described semi-quantitatively. The aim is to achieve a qualitative understanding of the causal relationships in the circuit, with the voltage causing the electric current and electrical resistance merely affecting it (see Fig. 4).

The previously purely qualitative concept of electrical resistance is then expanded to include a microscopic model of resistance. The aim here is to give students a better understanding of various conduction processes based on the Drude model. Ideal conductors, for example, are explained by the fact that the atomic cores in such a material are arranged very uniformly and the electrons hence almost never collide with the atomic cores (see Fig. 5, left). The fact that resistors have a not negligible electrical resistance can be explained in the model, for example, by assuming that the atomic cores are not uniformly arranged in the material, which means that collisions between the moving electrons and the atomic cores will frequently occur (see Fig. 5, right).

Fig. 3 Simple electric circuit with a light bulb

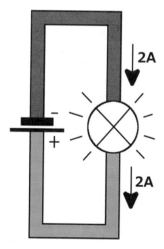

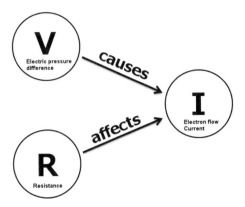

Fig. 4 Qualitative relationship of *V*, *I* and *R* in electric circuits

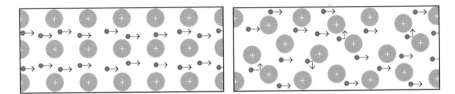

Fig. 5 Microscopic model of an ideal conductor (left) and a resistor (right)

2.4 *Parallel Circuits*

At the example of parallel circuits, students not only learn to clearly distinguish between the electric pressure concept and the electric current, but also that an (ideal) battery is a source of constant voltage and not constant current. In order to work out the electric current flowing through the different branches of a parallel circuit, students simply need to look at the electric pressure differences at the light bulbs and their electrical resistance. Based on the assumption that a big electric pressure difference causes an electric current of 2 A through a light bulb with a small electrical resistance and an electric current of 1 A through a light bulb with a big electrical resistance, students then simply need to add up the current through the different branches to get the current that needs to be supplied by the battery (see Fig. 6). By looking at electric pressure differences across the light bulbs first and then working out the electric current resulting from these pressure differences, this approach helps to make voltage—and not the electric current—the students' primary concept when analysing electric circuits. The colour coding also helps students to determine which light bulbs are connected in parallel as they simply need to compare the colours: If two bulbs have the same adjacent colours, they are connected in parallel (Fig. 6).

Fig. 6 Parallel circuit with three light bulbs. The right bulb has a higher electrical resistance than the other two

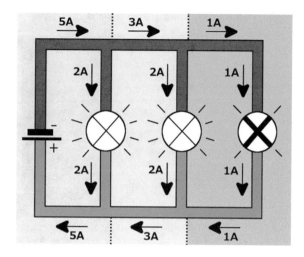

2.5 *Capacitors*

The analysis of capacitor charging and discharging using the model of "electric pressure" is supposed to help students understand the concept of transient states and dynamic model thinking, which they need for the analysis of series circuits. Transient states are introduced as it takes some time for the electric pressure in the different parts of the circuit to reach a steady state. In other words, the steady state is only achieved gradually over so-called transient states. By looking at the charging of a capacitor, the idea behind transient states becomes clearer.

In the very first moment, when the electric circuit gets connected to the battery, the battery causes a high electric pressure in section A and a low electric pressure in section C. In sections B and D, however, we still have a normal electric pressure since no electrons have flown through the light bulbs just yet (Fig. 7). During this transient state, electrons flow from section A through the top light bulb into section B, thereby increasing the electric pressure in section B. At the same time, we have electrons flowing from section D into section C, thereby decreasing the electric pressure in section D. As time goes by, the electric pressure in section B will increase until it aligns with the high electric pressure in section A and the electric pressure in section D will decrease until it aligns with the low electric pressure in section C. At that point, the steady state has been reached as the electric pressure in section B and D will not change anymore. The analysis described here also helps students overcome a number of common alternative conceptions, such as the belief that there are initially no electrons in the wires that electrons are stored inside the battery as oil is stored in an oil barrel or that the electric current flows sequentially from the negative terminal to the positive terminal.

Fig. 7 Transient state during
capacitor charging

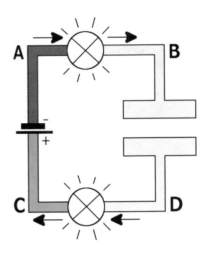

2.6 Series Circuits

As a next step, the concept of transient states is applied to a series circuit with a light
bulb with a high electrical resistance and a light bulb with a low electrical resistance
(see Fig. 8). At the beginning, when the electric circuit has not been connected to the
battery yet (initial state), we have a normal electric pressure in all parts of the circuit
(yellow). Once the battery is connected to the circuit (transient state), it creates a
high electric pressure in the top wire (red) and low electric pressure in the bottom
wire (blue). Since no electrons have flown through the light bulbs at that point, the

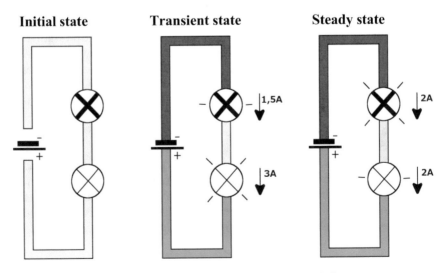

Fig. 8 Step by step analysis of a series circuit with two different light bulbs

Fig. 9 Transition from a qualitative to a quantitative relationship of V, R and I

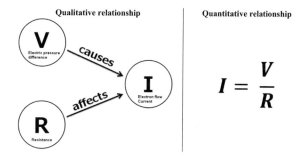

electric pressure in the middle part of the wire remains unchanged (yellow). In that transient state, we have the same electric pressure difference (= voltage) across both light bulbs. However, since the top light bulb has a higher resistance than the bottom light bulb, fewer electrons flow into the middle wire than out of it. Consequently, the electric pressure in the middle part of the wire decreases. As the pressure decreases, the electric pressure difference across the top light bulb increases, while the electric pressure difference across the bottom light bulb decreases, with the effect that the electric current through the top and bottom light bulb will align over time. Once the electric current through the top and the bottom light bulb are identical, the electric pressure in the middle part of the wire will not change anymore and the steady state has been reached.

2.7 Quantitative Relationship

Once a qualitative understanding of "voltage", "current" and "resistance" and their mutual relationship in simple electric circuits is established, the teaching concept aims to transform this qualitative understanding into an understanding for the quantitative relationship $I = \frac{V}{R}$ (Fig. 9). A more detailed description of the teaching concept and its theoretical background can be found in Burde (2018).

3 Quantitative Evaluation

3.1 Method and Sample

The purpose of the quantitative evaluation of the new curriculum was to find out whether it leads to a higher learning gain than traditional approaches to teaching electric circuits. In order to answer that question, a quasi-experimental field study was conducted with $N = 790$ students from Frankfurt/Main, Germany, taking part in the evaluation. The field study followed a pretest-posttest-control-group design, where the control group (CG) was taught the traditional way by 11 teachers for an average of 23.5 lessons (SD = 11.9) and the slightly larger experimental group (EG) was taught according to the new curriculum by 14 teachers for an average of 24.3 lessons (SD = 9.8). The CG consisted of 17 junior high school classes with a total of $N = 357$ students and the EG consisted of 19 junior high school classes with a total of $N = 433$ students. Regarding the group size and the number of lessons taught, both groups were, thus, comparable. Furthermore, the topic of electric circuits was covered in both groups for the first time.

3.2   Test Instrument

In order to evaluate the students' conceptual understanding of electric circuits, the same valid and reliable two-tier multiple-choice test was used in both groups for the pre- and posttest. This test instrument was developed in Vienna by an independent research group without reference to the new teaching concept and contained 22 items in its original form (Urban-Woldron and Hopf 2012). Since the original test instrument with its 22 items primarily focused on the concepts of current and resistance, but not on voltage, we extended the original test by another four items that evaluated the students' conceptual understanding of voltage. Against the background that the new teaching concept primarily aims to give students a better conceptual understanding of voltage while only four out of 26 items of the test instrument focus on voltage, it can be assumed that the test instrument is unbiased towards the new teaching concept.

The two-tier structure of the diagnostic multiple-choice test provides deeper insight into students' reasoning about circuits as they do not only have to answer questions (first tier), but also give an explanation to their answer (second tier). By analysing the combination of answer and explanation, it is not only possible to identify false-positive answers (i.e. correct answers with an inadequate explanation), but also typical alternative conceptions about electric circuits. An item was only counted as correct if the answer (first tier) as well as the explanation (second tier) was given correctly. Since there were 26 items in total, the highest achievable score in the multiple-choice test is therefore 26 points.

3.3   Empirical Results

As the students were not taught independently of each other but were grouped in different classes, their learning success depends heavily upon their class membership. In the terminology of multi-level analyses (MLA), the students are "nested" in school classes. In order to appropriately account for this hierarchical data structure of the sample, a multi-level analysis was conducted. Such a multi-level analysis, also called hierarchical linear model (HLM), provides the most adequate estimate of the net effect of the treatment and its statistical uncertainty (Fig. 10).

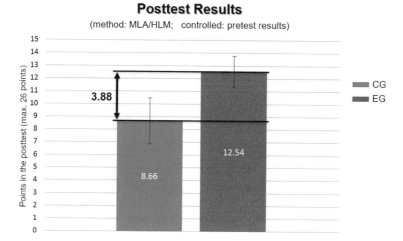

Fig. 10 Posttest results of the CG and EG

According to the HLM, the net effect of instruction is 3.88 points, which is a highly significant result and corresponds to a large effect of $d = .94$. It can, therefore, be assumed that the new teaching concept leads to a higher learning gain than traditional approaches to teaching electric circuits. Given the fact that the teachers of the experimental group taught their classes according to the new teaching concept for the first time and did not receive any in-service training, this is a remarkable result.

Thanks to the two-tier structure of the test instrument, it was also possible to analyse the students' alternative conceptions after instruction by means of a binary logistic multi-level analysis. In general, the new curriculum seems to lead to a better conceptual understanding as students of the EG either have a comparable, or significantly lower, probability to hold typical alternative conceptions after instruction than traditionally taught students. The new curriculum particularly appears to lead to a better understanding that voltage, in contrast to current, can only be measured between two points in a circuit because voltage refers to a difference in electric potential. A more detailed analysis of the empirical evaluation of the teaching concept can be found in Burde (2018).

4 Qualitative Evaluation

4.1 Method and Sample

Since the study was designed as a Design-Based-Research (DBR) project, it is not only of interest whether students achieve higher learning gains as a result of the new teaching concept, but also what the teachers think of it based on their practical experience. In order to get an idea of the teachers' perspective on the teaching concept, an online questionnaire was created and distributed to the 14 participating teachers by e-mail. The questionnaire had a length of about 30 min and was divided into different sections. The primary goal was to find out where the teachers see the strengths and the weaknesses of the teaching concept and whether they plan to teach according to it again in future.

4.2   The Teachers' Perspective

All teachers considered the introduction of voltage as potential difference to be a good idea and said that the air pressure analogy in combination with colour coding leads to a better conceptual understanding of electric circuits. In particular, the teachers said that the relationship between voltage and current becomes clearer thanks to the new teaching concept. They also praised the concept for its good teaching resources and the fact that the students get a first microscopic model of the processes in electric circuits. It has been criticised, however, that the choice of colours in the colour coding scheme contradicts the convention commonly used in physics, according to which the positive terminal is usually coloured red and the negative terminal blue. Another point of criticism was that particularly younger students struggled with the concept of transient states to explain series circuits.

Overall, however, the teachers perceived the teaching concept presented in this paper as a significant improvement, which is why 12 of the 14 participating teachers state that they plan to teach according to it in future. From the perspective of design-researchers, this is a very positive result, since the acceptance of the teaching concept by practitioners is a necessary precondition for a broad adoption in schools. For a comprehensive description of the teachers' perspective, please refer to Burde (2018).

5   Outlook: The EPo-EKo Project

Based on the research findings of the project presented in this paper, the joint Design-Based Research project EPo-EKo ["Electricity with Potential and Electricity with Contexts" (spelled with a "K" in German)] is currently being carried out by three German and two Austrian universities. Among other goals, the project aims to find out whether the significantly better conceptual understanding of students instructed according to the new teaching concept can be replicated with a bigger sample of teachers and students. Furthermore, it is planned to investigate the effects of context-based materials on the students' interest, self-concept and conceptual understanding when teaching electric circuits. Additionally, the project aims to shed a light on the teachers' pedagogical content knowledge (PCK) and beliefs about teaching and learning introductory electricity and how they change due to the implementation of new teaching materials. A more detailed description of the EPo-EKo project can be found in Haagen-Schützenhöfer, Burde, Hopf, Spatz & Wilhelm (2019).

References

Burde J-P (2018) Konzeption und Evaluation eines Unterrichtskonzepts zu einfachen Stromkreisen auf Basis des Elektronengasmodells. Logos, Berlin

Burde J-P, Wilhelm T (2016) Moment mal... (22) PdN-PidS 65(1):46–49

Burde J-P, Wilhelm T (2018) PERC proceedings 2017

Cohen R, Eylon B, Ganiel U (1983) Am J Phys 51(5):407–412

diSessa AA (1993) Cogn Instr 10(2–3):105–225

diSessa AA (2008) In: Vosniadou S (eds) International handbook of research on conceptual change. Routledge, London, pp 35–60

Duit R, Jung W, Rhöneck C (eds) (1985) Aspects of understanding electricity—proceedings of an international workshop (Schmidt & Klaunig, Kiel)

Gleixner C (1998) Einleuchtende Elektrizitätslehre mit Potenzial. LMU, München

Haagen-Schützenhöfer C, Burde J-P, Hopf M, Spatz V, Wilhelm T (2019) Using the electron-gas-model in lower secundary schools – a binational design-based research project. In: McLoughlin E, von Kampen P (eds) Concepts, strategies and models to enhance physics teaching and learning. Springer, Switzerland

Härtel H (2012) PdN-PidS 61(5):25–31

Herrmann F, Schmälzle P (1984) Der math und naturw Unterricht 37(8):476–482

Muckenfuß H, Walz A (1997) Neue Wege im Elektrikunterricht. Aulis Deubner, Köln

Rhöneck C (1986) Naturwissenschaften im Unterricht Physik 34(13):10–14

Schumacher M, Wiesner H (1997) DPG-Tagungsband 1996:573–578

Schwedes H, Schilling P (1983) Physica Didactica 10:159–170

Schwedes H, Dudeck W-G, Seibel C (1995) PdN-PidS 44(2):28–36

Steinberg MS, Wainwright CL (1993) TPT 31(6):353–357

Urban-Woldron H, Hopf M (2012) ZfDN 18:201–227

Waltner C, Späth S, Koller D, Wiesner H (2009) In: Höttecke D (ed) GDCP Dresden 2009, vol 30. Lit-Verlag, Münster, pp 182–84

Wilhelm T, Hopf M (2018) In: Schecker H, Wilhelm T, Hopf M, Duit R (eds) Schülervorstellungen und Physikunterricht (Springer-Spektrum, Wiesbaden), pp 115–38

Teaching Particle-Wave Duality with Double-Slit Single-Photon Interference in Dutch Secondary Schools

Ed van den Berg, Aernout van Rossum, Jeroen Grijsen, Henk Pol, and Jan van der Veen

Abstract A lesson series on quantum physics was taught to grade 12 students in 5 different secondary schools. Among others, the series contained a special demonstration experiment for particle-wave duality (double-slit single-photon interference) which was supplemented by the PhET applet quantum wave interference. Preconceptions on particles and waves were assessed in a pretest. Interviews were conducted with 24 students in 4 schools after the demonstration. A posttest was administered to 112 students in 3 schools. Some lessons were observed by an outside observer and/or videotaped, followed by interviews. Results indicate that learning objectives of quantum physics are well within reach of the 12th grade students. This shows especially in student discussions of the PhET applet. However, understanding of wave-particle duality is hindered by insufficient understanding of the prerequisite wave concept that waves are spread out in space. Most students associated a wave with an oscillation and somehow forgot about the spatial aspects. As a consequence, the "strangeness" of wave-particle duality was underappreciated. Also, the link with the de Broglie wavelength discussed early in the lesson series was not made. As in the 5th school quantum physics started several weeks later, we were able to try an improved lesson design for the demonstration successfully. Interviews are very helpful in catching misinterpretations of demonstrations and generating suggestions for improvement. Students do appreciate the confusing wonders of quantum physics but in the post-interviews seem unaware of the many quantum-based devices around them and the many everyday phenomena (like color) which need quantum physics for explanation. Perhaps as teachers and curriculum developers, we are too eager to overemphasize the exotic and strange aspects of quantum theory and underemphasize utility in everyday applications.

E. van den Berg (✉) · A. van Rossum · J. Grijsen · H. Pol · J. van der Veen
University of Twente, Drienerlolaan 5, Enschede 7522, NB, The Netherlands
e-mail: e.vandenberg-1@utwente.nl

J. Guisasola and K. Zuza (eds.), *Research and Innovation in Physics Education: Two Sides of the Same Coin*, Challenges in Physics Education,
https://doi.org/10.1007/978-3-030-51182-1_11

1 Introduction

In the Netherlands quantum physics was made compulsory for all students who are taking physics in the pre-university stream and has been included in the national final examinations since 2016. This concerns about 10% of the cohort of all 17/18-year old's, an annual total of about 20,000 students nationally. Prior to 2016 students studied aspects of modern physics such as the photo-electric effect and energy levels and spectra, but they did not encounter particle-wave duality, tunneling, wave functions, or the Heisenberg principle.

Quantum physics has surprised physicists from the start with stunning predictions and findings and spectacular success in producing powerful applications. However, it has taken a long time before quantum physics reached secondary education and only in the past 20 years quantum physics has gradually entered main stream secondary curricula and national examination programs. Nuffield A-level physics around 1970 did include some quantum physics and even contained a simple photon interference experiment at very low intensity, thus single-photon interference (Nuffield 1970). In the Netherlands, the pilot Project Modern Physics at eventually 40 schools (Hoekzema et al. 2007) paved the way for the inclusion of quantum physics in the national curriculum at all 500 schools which offer pre-university secondary education. The new curriculum emphasizes conceptual aspects of quantum physics rather than the mathematical formalism. For example, there is no Schrödinger equation and computations are limited to the simple particle-in-box model (Hoekzema et al. 2007).

'*Quantum mechanics leads to fundamental changes in the way the physical world is understood and how physical reality is perceived*' (Karakostas and Hadzidaki 2005) and to make students recognize this with some understanding is the challenge of quantum physics in secondary education. Many authors have documented learning difficulties with introductory quantum physics at the secondary and tertiary level, see Krijtenburg-Lewerissa et al. (2017). With regard to the double slit, Krijtenburg-Lewerissa et al. distinguished students with a pure classical view (light is always a wave, particles move in straight lines and cannot reach places where there is an obstacle in between), or some kind of quantum view in which they recognize that quantum objects have wave and particle properties but may still have difficulty with interference of single photons, or using the de Broglie wavelength to explain how double-slit interference changes as a function of mass and velocity of quantum particles. Wavefunctions can extend past classical boundaries so quantum objects can exist at positions which are not possible in classical physics.

We were eager to include laboratory/demonstration experiments; however, students interpret the information of demonstrations through their individual conceptual framework and misconceptions. What students see and think during observations may be quite different from what the teacher expects or imagines. Therefore, in this study, we checked on student observations and interpretations in order to find more effective ways of using demonstrations.

2 Research Questions

1. To what extent can students explain and apply the concept wave-particle duality after demonstration of the double slit?
2. Which conceptual problems may inhibit learning during the demonstration?
3. What are concrete suggestions to improve the learning effects of the demonstration?

3 Method

A team of teachers supported by the University of Twente developed supplementary materials including demonstrations on wave-particle duality (double-slit interference of single photons) and tunneling (microwaves) to support concept development, taking into account well-known learning difficulties in quantum physics (Krijtenburg-Lewerissa et al. 2017; Muller and Wiesner 2002). Teachers at 5 different schools used their own choice of textbook but inserted the supplementary materials including the emphasis on wave-particle duality and the double-slit demonstration and the PhET quantum wave interference applet. The textbooks at different schools differed in presentation but covered the same national syllabus topics. This evaluation focused on the double-slit demonstration and the concept of wave-particle duality. Evaluation data were collected through lesson observations, interviews immediately after the demonstration (1 school) or toward the end of the lesson series (3 schools) with a total of 24 students, interviews with 6 students were done in pairs while all other interviews were done individually (18 students), a pretest for prerequisite concepts and a different posttest with 112 students in 3 of the participating schools. At one school, the double-slit demonstration was observed live by the researcher in the classroom, at another school, it was video recorded. As quantum physics was scheduled earlier in the school year at some schools and later at other schools, interviews at an earlier school were used to revise the lesson for the double-slit demonstration, and this revision was tried out at the last school and results were checked in student interviews and the posttest.

4 The Demonstration

Figure 1 shows a set-up of a double-slit interference of single photons in a regular-sized suitcase which can be borrowed by schools. A laser beam passes through a filter which lets through only one out of 10^6 photons. With a laser of 5 mW and the slit only letting through about 15% of the remaining intensity, the distance between successive photons becomes about 12 cm and at any time there would be only 1 or 2 photons between slits and detector. A biconcave lens between slits and detector spreads the beam. The photons are then counted at 300 positions in the 1, 5 mm cross-section

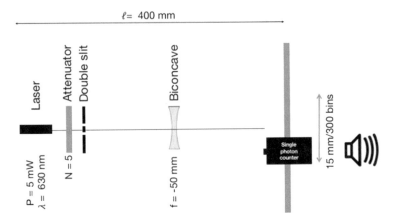

Fig. 1 Diagram of the single-photon interference experiment

of the beam. The counting is accompanied by sounds like those of a Geiger–Muller counter to underline the notion that photons are particles and arrive one-by-one. While counting, the *number of photons* versus *location* graph builds up on a screen (Fig. 3) and is projected on the wall. A detailed technical description is available from the authors. The demonstration was followed by a teacher presentation and discussion of the PhET quantum wave interference applet and concluded by the Dr. Quantum film. Three universities in the Netherlands now have a suitcase demonstration set available for use by schools (Fig. 2).

5   Results

Prerequisite concepts: Both written tests and interviews indicated that almost all students did not properly conceptualize the difference between waves and a particles, both before and after the lesson series. The most common student definition of a wave was a vibration or an oscillating object. Some defined a wave as a movement with a wavelength and a frequency and without mass. Some others defined a wave as many particles oscillating and so the wave-particle difference as many particles versus one. Few students defined a wave as a disturbance which propagates in space (3D) or on a surface (2D) and which *spreads out*. A particle is said to have mass and velocity and sometimes volume is mentioned. For wave-particle duality, we think the crucial difference between waves and particles is that waves are spread out and particles are localized with a definite volume. Duality does not make sense if this classical difference is not made explicit.

Lesson observations and interviews: The most striking result was that students at two schools in the demonstration lesson and in the subsequent interviews were not at all surprised with the outcomes of the double-slit experiment. They seemed

Fig. 2 Single-photon
interference in a suitcase

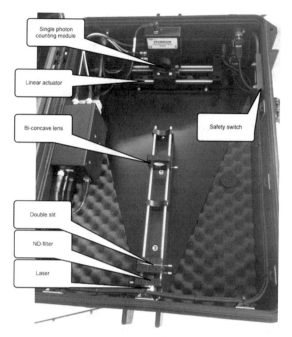

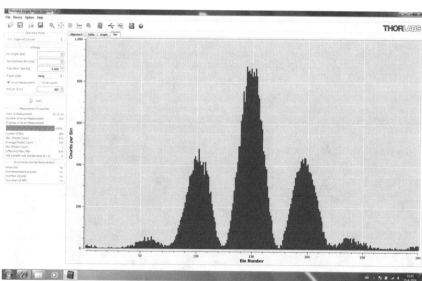

Fig. 3 Display of results

to think that physics and especially quantum physics always have strange outcomes, so why not. Some also seemed to have switched to the idea that all particles passing slits would exhibit interference, and even small bullets of paint would produce an interference pattern. Many students insufficiently realized the conflict and surprise that we physicists think they should experience. Their background in physical optics with the current Dutch Physics curriculum is too limited and they are used to physics being incomprehensible and counterintuitive anyway. Wave and particle? Photons or electrons interfering with themselves in a double-slit experiment? The interference pattern disappearing when a detector is put at one of the slits? O well, why not. Of course, one could react by "nice that twenty-first-century students are not surprised anymore by quantum theory", but we think that they insufficiently realize the meanings of wave, particle, probability distribution, and other classical and quantum concepts (Krijtenburg-Lewerissa et al. 2017; Muller and Wiesner 2002).

Questions about the PhET quantum wave interference applet (Fig. 4) were answered quite well. For example, that the electron could show up anywhere in the cloud if detected, that propagation was governed by wave phenomena and absorption in the screen by particle phenomena. Most students could also tell that the presence of a detector at one of the slits would result in disappearance of the interference pattern.

For try-out at the 5th school, we redesigned the double-slit demonstration lesson to articulate first the classical meaning of waves and particles and then conducted a series of double-slit experiments to build up to a final surprise. We tried out this

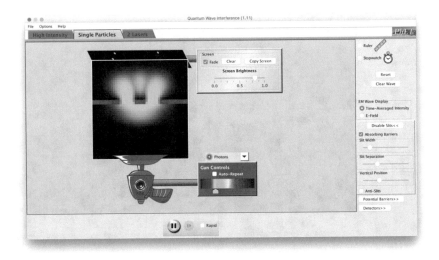

Fig. 4 Screenshot of photon passing double slit in the PhET quantum wave interference applet. *Image* from PhET Interactive Simulations, University of Colorado Boulder, licensed under CC BY 4.0

double-slit lesson in two parallel classes with a modified demonstration sequence and student sketches of predictions prior to each demo:

1. Spraying colored water (paint) through a double slit (several mm wide) using a common plant sprayer and observing the pattern;
2. A beam of parallel light rays through the same slit;
3. A video of water waves passing a double slit;
4. Laser light on a double slit;
5. Single-photon interference (the demo described above). In this, demo students also calculate the distance between photons from the power of the laser and the attenuation of the filter.

This turned out to be exactly the set of demonstrations in the well-known Dr. Quantum YouTube film. For each experiment, we asked students to sketch predictions for what they expected to see on the screen. This set-up led to more surprise and awareness of the strangeness of quantum results as seen in post-interviews at that school.

Right after the demonstration:

I: What have you just learned?
S1: About photons, and interference, and uh…
I: What can you tell me about photons?
S1: They have a particular energy and apparently, they interfere with each other and also individually.
I: Individually also, what do you mean by that… That they interfere individually?
S1: That was written on the board.

Further in the interview:

I: So then the sunglasses (= filter) were applied and then we got another graph, or something that looked the same?
S1: The same, isn't it?
I: Yes.
S1: I did not see another one.
I: No, no, the same. But they went one-by-one through the slit. Isn't that strange? They went one-by-one, and yet you get interference?
S1: Yes, apparently.
I: You did not find that strange? You think: well that is how it is?
S1: Well yeah…

What is meant by wave-particle duality? Question posed to two good students:

S2: I thought that this means that a particle exhibits either wave or particle behavior but not both at the same time. Or, I don't know.
S3: Yeah, it means that certain small particles are really waves as long as they are not observed. And as soon as they are observed they seem or have become particles and this duality means that they can be both but not at the same time.

S3: A wave is everywhere in space and a particle is at one spot. [comment: this was one of the very few students who got the essence of wave-particle difference]

Interviews after the improved version of the demonstration:

I: What can you tell me about that suitcase experiment?
S4: Yeah, they produced light particles and there was so much distance in between that you were sure the particles did not influence each other, some kilometers in between. Then with the photon detector they looked where most particles ended up and then you could see that there was an interference pattern.
I: Didn't you find that strange?
S4: Yeah, that was strange.

When showing the PhET applet:

I: While the photon propagates, can you tell where it is?
S4: No.
I: and what if you know where it ends up on the screen? Can you tell where it is before it hits the screen?
S4: No, I don't think so. At least I cannot tell.
I: Can you tell through which slit it went?
S4: Also not, no. You could say that if it ends up there [points], that it is more logical that it would go through the slit on the right, but I think that it is not a matter of logic in this way.
I: So if I ask you does it go through one slit or the other, what do you say?
S4: I do not answer that, I would not know. It could even go through both.

Furthermore, in interviews, many students seem to think that quantum theory is still very new and uncertain. When asked whether they could point to devices in the classroom where quantum theory had been applied, many students could not answer. They did not realize that anything containing semi-conductors like a laptop or mobile phone had required application of quantum calculations for the design and production of the chips. Even sunscreen oils and textile dyes may have been designed with quantum computations for wavelengths to be absorbed or reflected. Even viewing color with our eyes is a quantum process. Textbook and classroom explanations apparently had over emphasized the exotic aspects of quantum theory and underemphasized utility in the everyday environment.

Written posttest: In interviews, students had trouble remembering de Broglie wavelength from the first quantum lessons and linking it to the newly acquired knowledge about double-slit interference. In the posttest (112 students), they were asked what would change in the interference pattern if the velocity of electrons would be increased, 56% answer *nothing* and 25% correctly predicted that interference maxima/minima would be closer together. Increasing the mass but keeping the same velocity gave better results, 56% correctly predicted that the interference maxima would be closer together. Asking what would change if photons would go one-by-one through the slits, 86% *correctly* answered nothing, very likely influenced by the demonstration.

6 Conclusions

Secondary students in the pre-university track of Dutch secondary schools can learn to explain basic qualitative features of wave-particle duality and find this interesting. The use of the double-slit demonstration can contribute to their understanding of wave-particle duality if the conditions of prerequisite knowledge (classic wave vs particle difference) are met and demonstrations are conducted with attention to the proper sequence. The use of the PhET applet on quantum wave interference is very useful to visualize particle and wave properties and generate in-depth discussion. The interviews generated specific suggestions for modifying the demonstration lesson and a try-out confirmed their validity.

With all the challenges of teaching quantum physics, teachers should not neglect to point to concrete applications of quantum phenomena in everyday phenomena and technology such as the role of quantum tunneling in fusion of protons in the Sun, radioactive decay, and quantum phenomena in the many electronic devices.

With any demonstration, not just in quantum physics, teachers should be conscious of the possibility that students have observations and interpretations which deviate from those intended. The best way to find out and improve the didactical set-up is through interviews as shown in this study.

Acknowledgments With thanks to cooperating teachers (other than the authors) Timo Bomhof, Hajo Brandt, Sander Wenderich, and transcriber Imre van Veldhoven

References

Hoekzema DJ, van den Berg E, Schooten GJ, van Dijk L (2007) The particle/wave-in-a-box model in Dutch secondary schools. Phys Educ 42:391–398

Karakostas V, Hadzidaki P (2005) Realism vs constructivism in contemporary physics: the impact of the debate on the understanding of quantum theory and its instructional process. Sci Educ 14:607

Krijtenburg-Lewerissa, K, Pol H, Brinkman A, Joolingen W van (2017) Insights into teaching quantum mechanics in secondary and lower undergraduate education. Phys Rev ST Phys Educ Res 13:010109. https://doi.org/10.1103/physrevphyseducres.13.010109

Muller R, Wiesner H (2002) Teaching quantum mechanics on an introductory level. Am J Phys 70(3):200–209. https://doi.org/10.1119/1.1435346

Eye-Movement Study of Mechanics Problem Solving Using Multimodal Options

Jouni Viiri, Jarkko Kilpeläinen, Martina Kekule, Eizo Ohno, and Jarkko Hautala

Abstract We used an eye-tracking method to investigate students' approaches to solving a physics task using various representations. Eight upper-secondary school students from Finland took part in the study. We found that students who preferred either the text or graph representations watched the options differently, but they used both representations to be sure of their solution. Transitions between text and graph alternatives were different for students preferring either text or graph representations. Interviews revealed typical misconceptions about the concept of force. Implications for physics instruction are presented.

1 Introduction

In this study, we used an eye-tracking method to investigate students' problem-solving processes while they were completing a multiple-choice physics test. We were especially interested in how the strategies depended on whether students preferred text or graph representations of the multiple-choice alternatives. Eye-tracking studies base their interpretation of gaze allocation on the eye-mind hypothesis (Just and Carpenter 1980), according to which a person's attention is focused on

J. Viiri (✉)
Department of Teacher Education, University of Jyväskylä, Jyväskylä, Finland
e-mail: jouni.viiri@jyu.fi

J. Kilpeläinen
Department of Physics, University of Jyväskylä, Jyväskylä, Finland

M. Kekule
Department of Physics Education, Charles University, Prague, Czech Republic

E. Ohno
Faculty of Education, Hokkaido University, Sapporo, Japan

J. Hautala
Niilo Mäki Institute, Jyväskylä, Finland

© The Editor(s) (if applicable) and The Author(s), under exclusive license to Springer Nature Switzerland AG 2020
J. Guisasola and K. Zuza (eds.), *Research and Innovation in Physics Education: Two Sides of the Same Coin*, Challenges in Physics Education,
https://doi.org/10.1007/978-3-030-51182-1_12

the point of fixation. Therefore, eye movements have a spatiotemporal relationship to visual information, and gaze allocation provides indirect data regarding a person's cognitive process. In contrast to some studies showing the shortcomings of the eye-mind hypothesis (Hyönä 2010), other neurophysiological studies support the hypothesis (Kustov and Robinson 1996). For instance, in studies using multiple-choice tasks, attention and gaze are closely related (Holmqvist et al. 2015).

Recent studies of students' understanding of graphs, both with and without eye tracking (see (Susac et al. 2018) for review) shows that students have difficulty interpreting graphs. Studies using multiple-choice questions in which choices are represented in various ways (e.g., text of graph) show that students' gaze allocation seems to depend on the type of representation used (Viiri et al. 2017). Based on the dual representation paradigm, researchers have displayed two graph types at the same time and compared how different graph formats (line or bar) effect students' test results. They found that participants shifted their graph preference depending on task type and that participants used both graph types during the tasks to verify their answers (Strobel et al. 2016).

In this study, we similarly use the dual representation paradigm and explore students' strategies when multiple-choice alternatives are provided in both text and graph formats. The students' preferences for the representations are taken into account as well. Even though the multiple choices are represented differently, we assume that these representations are informationally equivalent in terms of answering the question. Two representations are termed informationally equivalent when they display the same relationships between the same objects (Strobel et al. 2016).

According to Mayer's cognitive theory of multimedia learning (Mayer 2003), humans process visual information via a visual channel and auditory information via a verbal channel. Also, humans often convert printed text into sound so that it can be processed in the verbal working memory. Therefore, we may consider students to be processing the text alternatives of the multiple-choice test in the verbal memory and graph alternatives in the visual memory. Prior research has focused on presenting multiple-choice alternatives in either text or graph form, and it is unknown whether providing these alternatives in both text and graph form in the same question side by side affects how students process items. Our aim is to investigate how students solve multiple-choice problems when alternatives are provided in two forms. Our research questions are as follows:

(1) How does students' gaze allocation differ depending on their graph/text preference?
(2) How do transitions between text and graph alternatives differ depending on students' graph/text preferences?
(3) How do students defend their choices?

2 Methods

2.1 Data Collection

The student participants were chosen from an upper-secondary school in Finland. They were completing the first and fifth physics courses (Opetushallitus 2003), both of which deal with mechanics. Their average age was 17 years. The students' parents signed a consent letter.

Before the eye-tracking test, students answered a nine-question multiple-choice pre-test to determine their conceptions of Newton's First and Second Laws. Based on the test results, a nine-student sample was chosen (very success full, success full or unsuccessful).

Eye-tracking data collection was performed during school days in spring of 2018. The eye-tracking device used in this study was the SMI RED250mobile. Prior to data collection, the eye-tracking unit was set to a 250 Hz sampling frequency, the fixation minimum period was 50 ms and the saccade was determined by an eye movement speed of at least 40°/s. The students completed the test independently on a computer. Before using the device, it was calibrated to determine the positions of the eyes of the participating student. The multiple-choice questions appeared on the screen one by one. Students selected the alternative they thought to be the correct answer with the mouse, after which a new question appeared on the screen, and they could not return to the previous tasks. Because of unstable eye track results, the data for one student were not used in this study. Ultimately, we had a sample size of eight students.

On the eye-tracking test, we had four qualitative items related to Newton's First and Second Laws. These items were based on the force concept inventory (FCI) test (Hestenes et al. 1992) and the representational variant of the force concept inventory (R-FCI) test (Nieminen et al. 2010). The layouts of the tasks were similar. The multiple-choice alternatives were presented in both the text and graph representations. For every item, the stem was presented in text form on the left side of the screen. Items were arranged in pairs related to same physics concepts, and they varied so that in one item, the text was on the left side and the graph on the right side, while in the other paired item, the order was reversed. Figure 3a depicts one item (a woman pushing a box) in which the text alternatives are on the left and the corresponding graph alternatives are on the right. Figure 3b shows a similar item (a man pushing a trolley) in which the text and graph alternatives are in reverse order as compared to the woman pushing a box item. The reason for this text/graph rotation was to determine whether students use the representation option (graph or text) they prefer or choose the alternatives that are closest to the stem. Otherwise, the layouts of the tasks were similar because there is evidence that different spatial layouts affect problem solvers' gaze movement (Holsanova et al. 2009). In all items, only one out of the five multiple-choice alternatives was correct, and the remaining alternatives were distractors related to typical student misconceptions.

Students' graph-text preferences were investigated with a five-point Likert scale questionnaire. The four questions used were "I am good at reading graphs", "I feel confident in reading graphs", "Generally, I prefer text form to graph form" and "Generally, I prefer graph form to text form". The test idea was adopted from (Strobel et al. 2016), in which the researchers used a similar short questionnaire to determine students' preferences for line or bar graphs. The graph-text preference measure was calculated as the difference between the preference scores for the questions.

After the eye-tracking tasks, students were interviewed. During these interviews, they were shown the eye-tracking items on a computer screen, and the interviewer asked, "Why did you choose that alternative?", "Which alternative did you immediately recognise as incorrect answer?" and "Did you have the second candidate, or did you hesitate to choose the answer you chose?" Other questions depended on a student's responses to the previous interview questions. After this, the students were shown their gaze plot videos. The students commented on their solution processes, and the interviewer asked, e.g., "What was your strategy in solving the task?", "Why didn't you look at that alternative at all?", "Why did you look so much at this specific alternative?" and "When you observed your record from the gaze plot video, was there anything surprising for you?".

2.2 Data Analysis

For the analysis of the eye-tracking data, we generated heat maps for each student and each task with SMI BeGaze software. Heat maps show how much a subject has viewed certain areas of the task, words or images when solving the problems. We created the areas of interest (AOIs) with the SMI Experiment Centre software programme. For every item, the stem and each multiple-choice alternative were a separate AOI (see Fig. 1). We created these AOIs to investigate the transitions between the AOIs and thus obtain information about the transitions between the stem and the various alternatives. Students did not see the AOIs at any stage of the test.

To analyse students' gaze transitions between various AOIs, we produced transition matrices for each student and each task (i.e. $8 \times 4 = 32$ transition matrices) based on students' fixations on the AOIs. The columns and the rows of the transition matrix represent the AOIs in the region, and the cells in the matrix indicate the number of transitions from the row AOI to the column AOI. For example, as seen in Fig. 2, a student performed three transitions from option (d) in graph representation to option (d) in text representation.

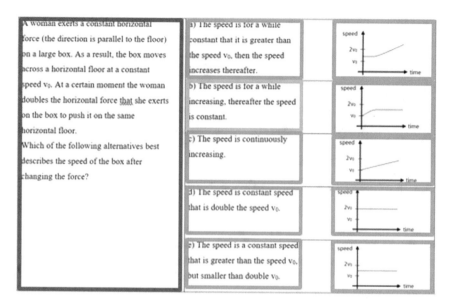

Fig. 1 AOIs of the task

p4	graph e	graph d	graph c	graph b	graph a	text e	text d	text c	text b	text a	stem
graph e						1					
graph d	1						3				
graph c		1									
graph b			1								
graph a				1							
text e		1									1
text d					1						1
text c											
text b											
text a											1
stem		2		1							

Fig. 2 Student #4 transition matrix for the "woman pushing a box" task

3 Results

3.1 Graph-Text Preference in Heat Maps

Heat maps (Fig. 3) show that Student #4, who prefers graphs generally, looked more at the graph alternatives independently, whether the graphs were on the right or left side on the screen. The student also looked at some text alternatives, but not as much as at the graph alternatives. For the "woman pushing a box" task, the student looked most at the text alternative (d), which he/she chose as his/her answer. In this task, the

a) woman pushing a box task　　　　　　　b) man trolley task

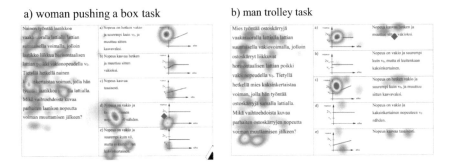

Fig. 3 Heat maps for two tasks completed by a student (#4) preferring graphs

student seems to have the typical misconception that if the force is doubled, the speed will also be twice as large. However, the "man trolley" task shows that the student's conception is not fixed because there, the student has chosen alternative (a), in which the speed is not doubled. The heat map shows that the student has almost discarded the double-speed alternative (d).

A text-oriented student's (#2) heat map for the "woman pushing a box" problem (Fig. 4) shows that he/she has mainly looked at the text alternatives. The student has chosen the incorrect alternative (e), in which the speed of the box is constant and larger than the original value. This student has also checked his or her "correct" thinking by looking at the corresponding graph alternative (e). Because the student has chosen the corresponding alternative in the "man pushing trolley" task, he/she seems to truly believe that although the force is double, the speed cannot be double, though it is larger than the initial speed.

a) woman pushing a box task　　　　　　　b) man trolley task

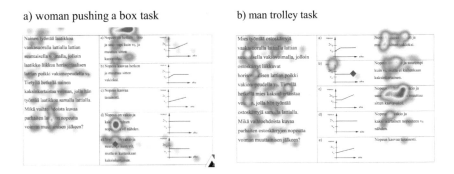

Fig. 4 Text-oriented student's (#2) heat maps for two similar tasks with different layouts

3.2   Graph-Text Preference in Transitions Between AOIs

We first examined the number of transitions from the stem to the options. This measure can be interpreted as the number of times a student has to reread the stem to connect it with the options and how difficult or easy it is for that student to remember the stem.

We found that in 80% of cases, there were, at maximum, seven transitions and that in 60% of cases, there was a maximum of five transitions. As we can see from Fig. 5, two students (5 and 8) showed more transitions from the stem to the options. Student #5 had many more, mostly for the "man pushing the trolley" and "woman pushing the box" tasks. This student also received the best results on the test. On the other hand, Student #8 received the worst results on the test. He solved none of the eye-tracking items correctly.

Secondly, we considered the transitions between different options, specifically from a text option to a graph option and vice versa. We also calculated the transitions within the option types, that is, from a text option to another text option and from a graph option to another graph option. Figure 6 shows the numbers of these three types of transitions for each student in all four tasks. The total number of transitions equals 100%. Transitions between the same kind of representation can suggest checking the options during the decision-making process, while transitions between different kinds of representations can suggest comparisons between representations.

For almost all tasks and all students, transitions between different representations are most common, and all students prefer these transitions for almost all tasks.

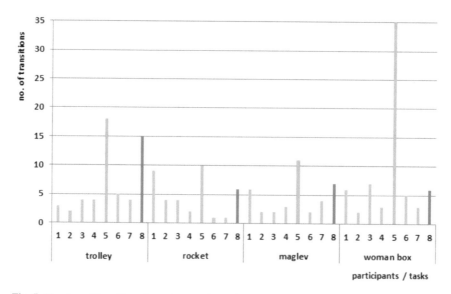

Fig. 5 Number of transitions from stem to alternatives

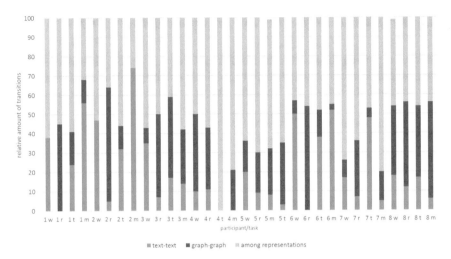

Fig. 6 Transitions between different representations (green), between graph options (red) and between text options (blue) in the tasks. (tasks: m = maglev, w = woman, t = trolley, r = rocket)

Students 1, 2 and 3 stated on the questionnaire that they preferred text representation. Accordingly, Students 1 and 2 seem to have more text-to-text transitions for the woman and maglev tasks. For the rocket task, they have more graph-graph transitions. Student 6 stated no preference, though he showed a similar text-preferring approach. Despite claiming to prefer text representation, Student 3 only has more text-to-text transitions during the woman task. Student 4 stated a graph preference, and for all tasks he indeed prefers either graph-graph transitions or text-graph transitions.

Table 1 shows the percentages of the different transition types for different student groups. From the table, we can see that students who report preferring text format make more transitions between text options than between graph options. The students who prefer graph representations make more transitions between the graph alternatives. In all groups, the transitions between different representation types are the most common. All student groups show similar levels of graph-graph transitions (about 25% of all transitions). However, the group of students who prefer graphs make far fewer text-text transitions and more transitions between different representation types. In contrast, students who prefer text make the fewest transitions between representation types.

Table 1 Percentage of transitions for students depending on their graph-text preferences

	Transitions		
	Text-text (%)	Graph-graph (%)	Between different representations (%)
Prefer text	29	22	49
Prefer graph	5	23	72
No preference	19	25	56

3.3 Students' Reasons for Their Answers

In the interview, students were asked the reasons for their choices. The answers revealed the students' typical non-Newtonian conceptions. They said, for instance, that "when the force is doubled, then the speed has to be doubled as well". They believe that a constant force acting on an object will lead to a constant speed for the object. Also, they believe that motion requires force. When giving evidence for their choices, they used everyday language, such as "I thought that there would be a kind of small kick when the motor starts and then it would be constant" or simply "it is reasonable".

Some students used explicitly correct ideas and physics concepts in their explanations. They could reason, e.g. "since there are no other forces in the space, when the motor pushes the rocket, the rocket speed increases all the time" or "Uhm, because there are no external forces acting in space, I thought that when it is pulling all the time, then the speed will also increase all the time". Another student said, "I thought that because no forces act on it, when more force is applied, the speed will increase because if there were an equal force on the opposite direction, it would continue moving with constant force, but because only one force is acting, I thought that its speed would increase".

The students were astonished to see from the gaze plot how they had been looking at the problem: "This is cool! From here, you can see where I stopped reading". They were mostly unaware of where they had allocated their attention during the task: "I was really astonished that when I was reading the stem, I also looked at the graphs many times" or "I was astonished that I looked at the graphs so little".

Some students could explain their transition behaviour, e.g. by saying that their problem-solving process involves considering various alternatives: "I first read the stem and started to go through the alternatives". Alternatively, they pick the best alternative and then check the corresponding graph/text option: "When reading the stem, I was already thinking about which might be the best alternative" or "By the end of the stem, I thought that the alternative that was constant would be my answer".

4 Conclusion

Our aim was to investigate students' gaze allocation on multiple-choice test when options are given both in text and graph form. Heat maps show that students' gaze allocation depends on their graph/text preferences. For instance, students who prefer graph representation tend to look more often at the graph alternatives than the text alternatives. Also, transitions between text and graph AOIs depend on students' graph/text preferences. Student's self-appraisal of their graph/text preferences may be wrong, but the eye-tracking results show their actual preferences. Often, students defended their choices by using non-Newtonian conceptions and everyday language, seldom using physics concepts and laws.

In this study, graph and text stimuli were placed side by side in the area of each option. This situation likely confused some students and provided a hint to others. These students are novices. They are not used to solving physics problems with graph and text options. Generally speaking, graph and text stimuli are equivalent informationally when one masters how to interpret them correctly. The arrangement forced students to consider that the graph as equivalent to the text and vice versa.

This was an explorative case study, and the results cannot be generalised. However, they do suggest new research directions. In future, we will collect additional data using the same questions and also questions addressing different physics areas to determine if there are any general patterns.

Teachers and researchers must remember students' preferences for graph and text representations. For some students, a graph alternative might be easier than a text alternative and vice versa. Students should be taught to combine both representation types and given opportunities to practise this.

References

Hestenes D, Wells M, Swackhamer G (1992) Phys Teach 30:141

Holmqvist K, Nyström M, Andersson R, Dewhurst R, Jarodzka H, van de Weijer J (2015) Eye tracking. Oxford University Press, Oxford

Holsanova J, Holmberg N, Holmqvist K (2009) Appl Cogn Psych 23:1215

Hyönä J (2010) Learn Instr 20:172

Just MA, Carpenter PA (1980) Psych Rev 87:329

Kustov AA, Robinson DL (1996) Nature 384:74

Mayer RE (2003) Cognitive theory of multimedia learning. In: The Cambridge handbook of multimedia learning, Cambridge University Press, New York

Nieminen P, Savinainen A, Viiri J (2010) Phys Rev Spec Top Phys Educ Res 6:020109

Opetushallitus (2003) Lukion opetussuunnitelman perusteet 2003 [Upper secondary school curriculum], Vammalan Kirjapaino Oy, Vammala

Strobel B, Saß S, Lindner MA, Köller O (2016) J Eye Mov Res 9:1

Susac A, Bubic A, Kazotti E, Planinic M, Palmovic M (2018) Phys Rev Phys Educ Res 14:020109

Viiri J, Kekule M, Isoniemi J, Hautala J (2017) In: Proceedings of the annual FMSERA symposium 2016 (Finnish Mathematics and Science Education Research Association, FMSERA), p 88

Derivatives, Integrals and Vectors in Introductory Mechanics: The Development of a Multi-representation Test for University Students

Marta Carli, Stefania Lippiello, Ornella Pantano, Mario Perona, and Giuseppe Tormen

Abstract The use of mathematical concepts and formal reasoning is one of the main hurdles for students entering introductory physics courses at university. The ability to apply mathematical tools in the context of physics also relies on the use of multiple representations, i.e., the different forms in which a concept can be expressed, such as words, graphs, numbers and formal language. Based on these considerations, we have developed a multiple-choice test consisting in 34 items aimed at investigating students' understanding of derivatives, integrals and vectors and their application in the context of introductory classical mechanics. The items were constructed using multiple representational formats and isomorphic items in mathematics, and in physics, in order to explore students' representational fluency and their ability to transfer knowledge and skills from mathematics to physics. The test has been administered to 1252 students enrolled in introductory courses at the University of Padova in Spring 2018. The results indicate that the test is a valid and reliable instrument and it provides interesting insight into students' difficulties in the use of mathematical concepts and methods in physics.

1 Introduction

The hurdles that students face as they enter physics courses at university are manifold. Among these difficulties, those regarding mathematics and formal reasoning are very common and a well-known issue among instructors and researchers.

M. Carli (✉) · O. Pantano · G. Tormen (Deceased)
Department of Physics and Astronomy, University of Padova, Via Marzolo 8, 35131 Padua, Italy
e-mail: marta.carli.1@unipd.it

S. Lippiello
Liceo Scientifico 'J. Da Ponte', Via S. Tommaso d'Aquino 12, 36061 Bassano del Grappa (VI), Italy

M. Perona
IIS 'C. Marchesi', Viale Codalunga 1, 35138 Padua, Italy

155

J. Guisasola and K. Zuza (eds.), *Research and Innovation in Physics Education: Two Sides of the Same Coin*, Challenges in Physics Education,
https://doi.org/10.1007/978-3-030-51182-1_13

Mathematics is often referred to as the backbone of physics (Bing and Redish 2009). In fact, learning to use mathematics effectively and efficiently has been recognized as one of the key dimensions for developing expertise in physics (Pantano and Cornet 2018). The concepts and methods of mathematics are used to model and describe physical phenomena, to make predictions about them, and to compute numerical results. However, knowing and applying mathematical concepts and methods in a purely mathematical context is not enough to use them effectively in a physical context. Using mathematics in science requires not only knowledge of mathematical tools and computational skills, but also less obvious processes such as the ability to map physics to mathematics, or vice versa, mathematics to physics (Bing and Redish 2009). The interpretation of the mathematical 'language' itself can be very different when used by physicists (Redish 2005; Redish and Kuo 2015). In fact, many researchers in physics have paid attention to the transfer of mathematical knowledge and skills to physics (Britton et al. 2005; Roberts et al. 2007), a competence that cannot be taken for granted.

On the other hand, when physics instructors describe students' difficulties, they usually categorize them according to the disciplinary content. However, research has shown that representational fluency, i.e., the competent use of different representations (graphs, words, diagrams, equations, etc.) to describe a concept, is a fundamental competence for the learning of physics and is typical of discipline experts (Van Heuvelen 1991; Redish 2003; Etkina et al. 2006).

Moving from this picture, this paper addresses the following question: how can we build an instrument that tests first year university students' difficulties in terms of (a) the transfer of knowledge and skills from mathematics to physics and (b) the use of different representations, in particular, for those mathematical topics that are relevant for introductory mechanics?

In order to answer this question, we have developed a distractor-driven multiple-choice test and we have administered it to students enrolled in the first year of 23 degree courses of the Schools of Science and of Engineering at the University of Padova. In this paper, after a description of the context and background of the research, the design and validation of the test are presented, together with an analysis of its reliability and the discussion of two relevant examples that can help exploring the two parts of the research question more in detail. The implications of these results for instruction are discussed in the last section.

2   Context and Background

The test was developed in the context of a project aimed at supporting both students and instructors of introductory physics courses at the University of Padova. The starting point was a survey conducted in June 2016, aimed at identifying the most common mistakes committed by first year students in their final physics exam. The respondents were 37 physics instructors, teaching introductory physics courses in the degree programs of the Schools of Science and of Engineering. The two topics that instructors recognized as a common source of difficulty for their students were 'vectors and trigonometry,' reported by 52% of the instructors, and 'basic calculus topics' (derivatives and integrals), pointed out by 43% of the instructors. Therefore, we decided to focus our research on these topics. In order to examine students' difficulties more in depth, we analyzed the final written exams of the same courses in academic years 2015/2016 and 2016/2017. This analysis confirmed the results of the survey. We then compared our students' difficulties with the results of PER literature.

There have been extensive studies on students' difficulties with the mathematical handling of the relationship between position, velocity and acceleration; in particular, it has been shown that students have considerable trouble understanding and using graphs in the context of kinematics. A first taxonomy of students' difficulties with kinematics graphs was compiled by McDermott et al. (McDermott et al. 1987). A decade later, Beichner (1994) designed and validated the Test of Understanding Graphs in Kinematics (TUG-K), consisting of 21 multiple-choice questions spanning seven typical students' difficulties; a modified version of the TUG-K was recently published by Zavala et al. (2017). Other authors explored some subtopics more in detail, with or without an explicit physical setting (Christensen and Thompson 2012; Wemyss and van Kampen 2013; Bollen et al. 2016; Nguyen and Rebello 2011). In a purely mathematical context, Epstein (2007, 2013) developed a concept test (calculus concept inventory) on the basic principles of differential calculus. More recently, Dominguez et al. (2017) proposed the Test of Understanding Graphs in Calculus (TUG-C), the counterpart of the TUG-K in the context of mathematics. Concerning students' ability to transfer calculus competences from mathematics to physics, particularly, relevant to our research were the works by researchers at the University of Zagreb (Planinic et al. 2012, 2013; Ivanjek et al. 2016).

Familiarity with vectors is essential to most topics covered by introductory physics courses and is fundamental to the study of Newtonian mechanics; however, many students enter their first year at university lacking the necessary working knowledge of vector concepts and methods. One of the first authors that systematically addressed this problem was Knight (1995), who developed the Vector Knowledge Test. More recently, Barniol and Zavala (2014) compiled a taxonomy of the most frequent errors made by university students when learning about vectors in introductory physics courses, upon which they based a multiple-choice test, the Test of Understanding of Vectors (TUV). Even for vectors, researchers have considered isomorphic problems in mathematics and physics in order to compare students' performance in the two contexts (Shaffer and McDermott 2005; Van Deventer and Wittmann 2007; Van Deventer 2008). Remarkably, in both contexts, students' difficulties can persist even after instruction (Nguyen and Meltzer 2003; Flores et al. 2004).

Representational fluency in solving physics problems has been addressed by many authors. Nieminen et al. (2010) have developed the representational variant of the force concept inventory (R-FCI), in which nine items from the FCI were redesigned using different representations. Some authors (Meltzer 2005; Kohl and Finkelstein 2005; De Cock 2012) have examined student performance in isomorphic problems formulated in different representational formats, finding statistically significant differences between different representations. More recently, Klein et al. (2017) have published a test of representational competence in kinematics (KiRC), with questions comprising different combinations of four representational formats.

Moving from this background, we designed a tool aimed at analyzing students' main difficulties in derivatives and integrals on one side, and vectors on the other, with an explicit focus on the transfer of mathematical knowledge and skills to physics and on the use of different representations.

3   Research Design and Methods

The research involved 23 degree programs at the University of Padova. Given the high number of students to be surveyed, we opted for a quantitative tool: a distractor-driven multiple-choice test.

One of the goals of the instrument was to explore students' representational fluency. We chose to distinguish four representations or 'languages':

- Verbal or 'natural' language, which we labeled as 'words' (W). Simple symbolic language was included in this category.
- Graphical language (G), comprising graphs and diagrams.
- Numbers (N), mainly consisting in numerical results.
- Formal mathematical languages (F), comprising algebraic expressions and equations.

Each item is characterized by the representational format of the question (input) and the multiple-choice answers (output). When the output language is different from the input language, students need to translate from one language to the other in order to answer the question. For each mathematical concept, several items were written considering different combinations of the four representations.

Since one of the main goals of our research was to evaluate the transfer of knowledge and skills from mathematics to physics, the instrument should contain corresponding items in the two contexts. Therefore, for each mathematical item, an isomorphic physics item was written. We chose mechanics as the physical context, since this topic is covered by all introductory physics courses at university and elements of mechanics are also taught in high school.

Initially, we developed two separate tests on the two topics (derivatives and integrals; vectors). Some of the items were adapted versions of items from existing tests. The distractors were designed starting from the results of our survey and from the literature. The test on derivatives and integrals contained 34 items, 12 in the context of mathematics and 22 in the context of physics (some mathematical items had more than one physics counterpart). The test on vectors contained 32 items, 13 in the context of mathematics and 19 in the context of physics. The two tests were checked by discipline experts (faculty members and physics instructors) for content validity and were then administered in March 2017 to 71 students enrolled in the degree program in architectural engineering. Both whole test and item reliability analysis were carried out, and semi-structured interviews were conducted in order to gather evidence for students' reasoning. Based on these results, several items and/or distractors were deleted, modified or rephrased. To achieve perfect isomorphism between the mathematics and the physics part, some items were modified while redundant questions were deleted. We also decided to consider more combinations of the input and output languages and we added some additional items accordingly.

In the final version of the test, the initial pool of items was reduced to 34 (17 items in the context of mathematics + 17 isomorphic items in the context of physics) according to Table 1. This set of items comprises the more relevant combinations of the four representational formats and satisfies the required isomorphism between mathematics and physics, while maintaining a reasonable total length of the test. The test was reviewed by three groups of experts for content validity: physics instructors, mathematics instructors and five independent researchers for representational categorization. Finally, the test was administered online via the Moodle platform in March 2018. Participation was voluntary but encouraged by instructors.

Table 1 Summary of test items

Item (math)	Topic (math)	Item (phys)	Topic (phys)	Representations
1	Derivatives	1F	Position → velocity	W → W
2	Derivatives	2F	Position → velocity	G → G
3	Derivatives	3F	Position → velocity	G → N
4	Derivatives	4F	Position → velocity	W → N
5	Derivatives	5F	Position → velocity	G → W
6	Derivatives	6F	Position → velocity	W → G
7	Integrals	7F	Velocity → displacement	G → W
8	Integrals	8F	Velocity → displacement	G → G
9	Integrals	9F	Velocity → displacement	G → N
10	Vectors (representation)	10F	Velocity (2D)	W → G
11	Vectors (representation)	11F	Velocity (2D)	G → F
12	Vectors (components)	12F	Forces	G → F
13	Vectors (sum-magnitude)	13F	Forces	G → F
14	Vectors (sum-components)	14F	Forces	G → F
15	Vectors (difference)	15F	Velocity (2D)	G → G
16	Vectors (dot product)	16F	Work	G/F → N/F
17	Vectors (cross-product)	17F	Torque	G/F → N/F

The label 'F' identifies items in the context of physics

4 Reliability and Instrument Characteristics

The test was taken by 1252 students enrolled in 23 different degree programs at the University of Padova. A mean test score of 5.8/10 was achieved for the whole sample, with a standard deviation of 2.4 and a standard error of the mean of 0.7. The mean score in the mathematics part (6.1) was significantly higher ($t = 13.39$, df $= 1251$, $p < 0.005$) than the mean score in the physics part (5.5) and the two distributions are correlated ($R = 0.79$). Calculation of the Kuder–Richardson index (KR20) as a measure of internal consistency of the test yielded is 0.91, suggesting that the test as a whole is reliable. The calculated Ferguson's delta of 0.99 suggests that the test has a good global discriminatory power too. Item facility indices (percent of correct answers) ranged from 0.32 (item 17F) to 0.94 (item 7), as reported in Fig. 1. The average was 0.58, corresponding to the mean test score normalized to one.

The average point-biserial coefficient (r_i) was 0.50. All items fulfill the criterion $r_i \geq 0.2$, suggesting that the test is reliable also at the single item level. As for the discriminatory index (D_i), we obtained an average value of 0.43; all items—except item 11—satisfy the desired criterion $D_i \geq 0.3$, meaning that the test discriminates reasonably well among students. Item 11, the only one that had a discriminatory index slightly below threshold (0.29), also had a high facility index (0.89), meaning that most students answered it correctly; it should be modified in a revised version of the test. A summary of the reliability results is reported in Table 2.

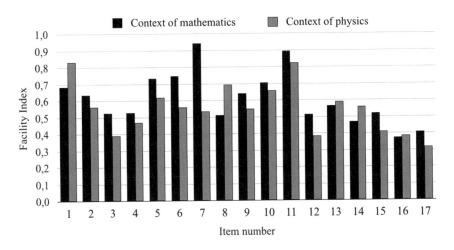

Fig. 1 Facility index versus item number

Test statistic	Desired value	Value
Kuder–Richardson index	>0.7	0.91
Ferguson's delta	>0.9	0.99
Facility index	[0.3, 0.9]	[0.32, 0.94], av. 0.58
Point-biserial coefficient	≥0.2	[0.29, 0.62], av. 0.50
Discriminatory index	≥0.3	[0.26, 0.58], av. 0.43

Table 2 Summary of statistical measures used for evaluating test reliability

5 Analysis of Two Relevant Cases

In the following, we will discuss two relevant cases that we found interesting for getting insight into the two foci of our research question: the transfer of knowledge and skills from mathematics to physics and representational fluency.

5.1 Mathematics and Physics

As an example of students' difficulties in transferring mathematical knowledge and skills to physics, we analyze and compare item 3 and its physics counterpart, item 3F. The two items are reported in Fig. 2. In item 3, students have to calculate the derivative at a given point quantitatively, given the graph of a function. In the physics version, students are asked to calculate the velocity of an object at a given instant, given the position-time graph of the object. Since the curve is a straight line in an interval around the given point (instant), the tangent to the curve is the line itself and its slope can be obtained by calculating $\Delta y/\Delta x$ ($\Delta s/\Delta t$) in a convenient Δx (Δt) interval.

This apparently simple problem has been widely discussed in literature. Students often have trouble determining slopes, especially when tangent lines do not go straight through the origin. Common mistakes consist in (1) computing the slope at a point by simply dividing y/x (as if it passed through the origin) and (2) confusing slope and height, thus reporting the y value as the slope. The distractors designed for our items reflect these commonly reported mistakes.

Students' responses for items 3 and 3F are reported in the pie charts in Fig. 2. 53% of the students answered item 3 correctly in a purely mathematical context. The most common wrong answer (17%) was distractor D, corresponding to the 'y/x' mistake. The second preferred distractor was B (13%), corresponding to the 'slope-height confusion' mistake. In the context of physics, the situation is very different. Only 39% of the students answered the question correctly, while the majority of students (45%) chose distractor D. Instead, distractor B was chosen by only 5% of

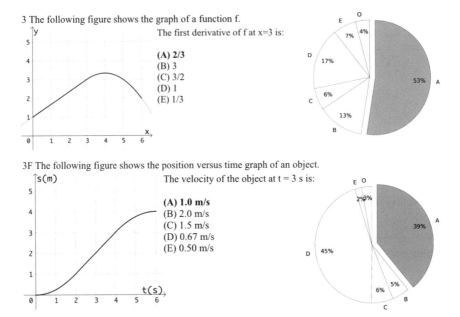

3 The following figure shows the graph of a function f.
The first derivative of f at x=3 is:

(A) 2/3
(B) 3
(C) 3/2
(D) 1
(E) 1/3

3F The following figure shows the position versus time graph of an object.
The velocity of the object at t = 3 s is:

(A) 1.0 m/s
(B) 2.0 m/s
(C) 1.5 m/s
(D) 0.67 m/s
(E) 0.50 m/s

Fig. 2 Items 3 and 3F. The correct answer is in bold (the order of options was randomized in the test administered to the students). Students' responses are reported in the pie charts

the students. These results suggest that the mathematical procedure is only part of the problem.

Students' interviews give us a hint about why they have so many difficulties in answering this question correctly in the context of physics. While solving item 3F during the interviews, many students defined velocity as 'space divided by time' and interpreted 'space' and 'time' as position and instant of time, respectively, which led them to the choice of distractor D. It is remarkable that this misconception appeared even when the mathematical strategy had been prompted by proposing an almost identical question (item 3) in the context of mathematics. In other words, incomplete transfer from mathematics to physics has occurred: some of the students that answered item 3 correctly did not recall the same mathematical procedure while solving item 3F, but switched to a different reasoning where they recalled some (incorrect) definitions of the physical quantities involved in the problem.

We got further interesting results by comparing students enrolled in different degree programs. Here, we compare three of them: mathematics, physics and architectural engineering. Students' responses to items 3 and 3F for these degree courses are reported in Fig. 3. The answers of students enrolled in physics and in mathematics reflect the distribution of the whole sample, even if with higher percentages of correct answers; the majority of students answered item 3F correctly, but distractor D was chosen by a higher number of students in item 3F than in item 3. Students enrolled in architectural engineering made different choices. In the context of mathematics

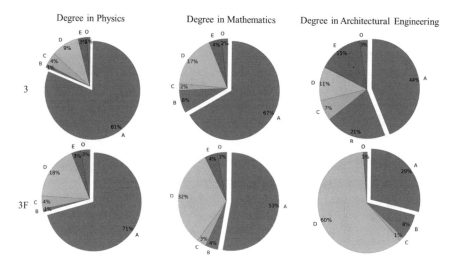

Fig. 3 Students' choices for item 3 and 3F for three different degree courses

(item 3), distractor B (21%) was the preferred distractor and E (15%) is the second choice, while D comes only third (11%). However, in the context of physics (item 3F), 60% of the students chose option D with little or no relationship with the distribution of answers in the context of mathematics, reinforcing the thesis that some students apply a different reasoning in the two contexts.

5.2 Representational Fluency

As an example of how switching between different representations can be an issue for students, we compare items 2F and 3F, both in the context of physics. Item 2F is reported in Fig. 4, while item 3F was reported in Fig. 2. The mathematical topic of item 2F is the same as item 3F (first derivative) and, in the same way, it is contextualized in physics by the relationship between position and velocity. For both items, the input is a position-time graph, but in item 2F students are asked to derive the velocity-time graph from the position-time graph, while in item 3F, they have to calculate velocity at a given instant. In other words, in item 2F, the output is given in the same language as the input, while in item 3F, students need to translate graphical information into numbers.

By looking at the pie charts in Figs. 2 and 4, we can compare students' performance in the two items. Item 2F was answered correctly by 56% of the students, and distractor C (a sort of 'shape maintaining linearization' of the position-time graph) was the most common wrong answer (17%). Instead, only 39% of the students answered item 3F correctly, with the majority of students (45%) choosing distractor D, as discussed above. Some reasons for this high mistake rate in item 3F have already

2F The following figure shows the position versus time graph of an object.

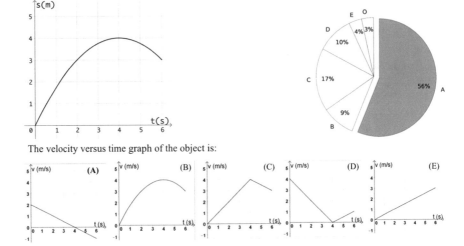

The velocity versus time graph of the object is:

Fig. 4 Item 2F. The correct answer is in bold (the order of options was randomized in the test administered to the students). Students' responses for this item are reported in the pie chart

been considered in the previous section, but difficulties could also arise from issues with representational fluency. These difficulties were confirmed by students' interviews. For instance, item 2F was answered correctly by a student that we will label as S2, who answered item 3F incorrectly; instead, we provide parts of the interview transcript to gain deeper insight into her reasoning.

I: "This is item 2F. You solved it correctly in the test. Can you explain your reasoning?"

S2: "Yes. Velocity is the first derivative [of position]. If there is a maximum here [indicates the position-time graph], then [the velocity] is zero, so the answer is either A or D or E. Here [the position-time graph] goes up, so the velocity is positive, and here it goes down, so the velocity is negative. Ok, it's A."

I: "Correct. Now, let's have a look at item 3F."

S2: "Here we go. I had trouble in the items that asked: 'how much is...'. Well.. shall I compute the area? If so, I need an integral."

M: "Why the area?"

S2: "Mmm, indeed I am not sure. So, maybe I should make a simple calculation, I mean velocity is space divided by time, it's... 2/3, so... Mmm, not sure".

I: "What is not convincing you?"

S2: "I think velocity is space divided by time only if it is constant, but here it is moving... I mean it is not constant".

M: "So when it is not constant, what do you have to do?".

S2: "I thought it had something do to with the integral, but it doesn't make sense. I think I have to take the derivative. But the derivative of what? I expected x or x to the second… I understand it must be the same as the previous question [item 2F], but I am confused.

I: *"You solved item 2F correctly What is confusing you here?"*

S2: *"That one [item 2F] was qualitative. To be more specific [quantitative] I need an equation".*

The reasoning of S2 while solving item 2F suggests that she has some understanding of the relationship between the graph of a function and the graph of its derivative, and that she can transfer the reasoning to a physical context as long as only graphs are involved. Although the proposed graphs were not only qualitative (numbers were reported on the axes), S2 is right when she says that she can solve item 2F by using only qualitative considerations on the trend of the position-time graph, thus involving only visuospatial processes and the declarative knowledge of the relationship between the trend of a graph and the sign of the derivative of the corresponding function. However, S2 fails to calculate a simple point derivative in a linear portion of a graph, claiming that she needs an equation (formal language) to calculate numbers. This student seems to lack representational fluency, limiting her working knowledge of concepts to isolated (sets of) representations without being able to translate from, for instance, graphs to numbers. During her discussion of item 3F, S2 also made some other common mistakes related to the problem. At some point, she refers to velocity as 'space divided by time,' although she corrects herself when encouraged to do so. She also displays some confusion between the concepts of integral/area and derivative/slope, thus committing a mistake known as 'slope/area confusion.' Remarkably, these mistakes and uncertainties appear only when S2 is asked to calculate numerical information from the graph, whereas in the graph-only version of the item, she is confident about the interpretation of velocity as the first derivative of position.

To get further insight into students' competence in the different representations, we calculated partial scores for each representational format by considering only items that contained that representation. We then compared the performance of individual students in the whole test, in the two disciplinary contexts, and in each representational format. We found some interesting situations. A remarkable example is that of three students who got the same score in the whole test (3.2/10) and also in subsets of it, namely in each of the two disciplinary contexts (mathematics: 3.5 and physics: 2.9) and in the two main topics (derivatives and integrals: 3.3, vectors: 3.1). However, we got completely different profiles when we compared their representational scores, as reported in Table 3.

Table 3 An example of three students who got the same score in the whole test (3.2/10) and on a disciplinary basis, but different partial scores in the four representational formats

Representational format	Student A	Student B	Student C
Words	4.8	4.3	2.8
Graphs	2.8	2.8	3.2
Formal language	3.3	2.5	4.3
Numbers	0.9	3.6	2.3

Student A performed much better in questions involving words than in those containing other representations, and he got a very low score in questions involving numbers. Student B did better than his average in questions containing words or numbers, but he did worse when graphs or equations were involved. On the contrary, student C did much better in items involving equations than items containing other representational formats. These results are interesting since these differences are usually not detected by traditional tests that only give feedback at a global or topic level and could therefore be useful for students to plan their own empowerment in a more personalized way.

6 Conclusions and Implications

The description of students' difficulties when they enter a physics course in their first year at university is complex. However, we believe that this instrument contributes to getting deeper insight into some of these difficulties by focusing on two competences that are often underestimated by both instructors and students: the transfer of knowledge and skills from mathematics to physics and representational fluency.

The relationship between mathematics and physics is a delicate issue in university instruction, where the two disciplines are usually taught separately, and instructors come from different departments and backgrounds. The ability to use mathematical tools and methods in the context of physics is a competence that is often overlooked by both mathematicians and physicists. We hope that our results help instructors becoming more aware of the fact that this transfer is not straightforward, and the difficulties in the use of mathematical reasoning and tools in physics may persist even when a mathematical concept has been taught and understood in a purely mathematical context.

On the other hand, since representational fluency is usually internalized by expert solvers, many instructors at the university level might take it for granted. However, first year students are basically novice solvers and they need explicit training on it. Being aware of the different representations and of the different individual abilities in their use can be a first step toward the activation of strategies that eventually produce an improvement of the teaching and learning of physics.

Acknowledgements This work is part of the *FisicaMente* project, jointly funded by the University of Padova and the *PLS—Piano Lauree Scientifiche* (Scientific Degrees Plan) of the Italian Ministry of Education, University and Research (MIUR). We acknowledge the School of Science and the School of Engineering of the University of Padova for both financial and motivational support. We also thank all the students and instructors who volunteered to take or administer the test, and all the colleagues who provided helpful comments. A special thanks goes to Dott. Francesca Zanandrea for her help with data processing.

References

Barniol P, Zavala G (2014) Test of understanding of vectors: A reliable multiple-choice vector concept test. Phys Rev ST Phys Educ Res 10:010121

Beichner RJ (1994) Testing student interpretation of kinematics graphs. Am J Phys 62(8):750–762

Bing TJ, Redish EF (2009) Analyzing problem solving using math in physics: epistemological framing via warrants. Phys Rev Spec Top Phys Educ Res 5(2):020108

Bollen L, De Cock M, Zuza K, Guisasola J, van Kampen P (2016) Generalizing a categorization of students' interpretations of linear kinematics graphs. Phys Rev Phys Educ Res 12:010108

Britton S, New PB, Sharma MD, Yardley D (2005) A case study of the transfer of mathematics skills by university students. Int J Math Educ Sci Technol 36(1):1–13

Christensen WM, Thompson JR (2012) Investigating graphical representations of slope and derivative without a physics context. Phys Rev ST Phys Educ Res 8(2):023101

De Cock M (2012) Representation use and strategy choice in physics problem solving. Phys Rev ST Phys Educ Res 8:020117

Dominguez A, Barniol P, Zavala G (2017) Test of understanding graphs in calculus: test of students' interpretation of calculus graphs. Eurasia J Math Sci Technol Educ 13:6507–6531

Epstein J (2007) Development and validation of the calculus concept inventory. Paper presented at the ninth international conference on mathematics education in a global community. Charlotte, NC

Epstein J (2013) The calculus concept inventory—measurement of the effect of teaching methodology in mathematics. Not Am Math Soc 60(8):1018–1026

Etkina E, Van Heuvelen A, White-Brahmia S, Brookes DT, Gentile M, Murthy S, Rosengrant D, Warren A (2006) Scientific abilities and their assessment. Phys Rev ST Phys Educ Res 2:020103

Flores S, Kanim SE, Kautz CH (2004) Student use of vectors in introductory mechanics. Am J Phys 72(4):460–468

Ivanjek L, Susac A, Planinic M, Andrasevic A, Milin-Sipus Z (2016) Student reasoning about graphs in different contexts. Phys Rev Phys Educ Res 12:010106

Klein P, Müller A, Kuhn J (2017) Assessment of representational competence in kinematics. Phys Rev Phys Educ Res 13:010132

Knight RD (1995) The vector knowledge of beginning physics students. Phys Teach 33(2):74–77

Kohl PB, Finkelstein ND (2005) Student representational competence and self-assessment when solving physics problems. Phys Rev ST Phys Educ Res 1:010104

McDermott LC, Rosenquist ML, van Zee EH (1987) Student difficulties in connecting graphs and physics: examples from kinematics. Am J Phys 55:503–513

Meltzer D (2005) Relation between students' problem-solving performance and representational format. Am J Phys 73(5):463–478

Nguyen N, Meltzer DE (2003) Initial understanding of vector concepts among students in introductory physics course. Am J Phys 71(6):630–638

Nguyen D-H, Rebello NS (2011) Students' understanding and application of the area under the curve concept in physics problems. Phys Rev ST Phys Educ Res 7:010112

Nieminen P, Savinainen A, Viiri J (2010) Force concept Inventory-based multiple-choice test for investigating students' representational consistency. Phys Rev ST Phys Educ Res 6:020109

Pantano O, Cornet F (eds) (2018) Tuning guidelines and reference points for the design and delivery of degree programmes in physics. University of Groningen. Published in the framework of the CALOHEE Project 2016–2018 (Agreement number: 562148-EPP-1-2015-1-NL-EPPKA3-PIFORWARD) funded with support from the European Commission, https://www.calohee.eu

Planinic M, Milin-Sipus Z, Katic H, Susac A, Ivanjek L (2012) Comparison of student understanding of line graph slope in physics and mathematics. Int J Sci Math Educ 10:1393–1414

Planinic M, Ivanjek L, Susac A, Milin-Sipus Z (2013) Comparison of university students' understanding of graphs in different contexts. Phys Rev ST Phys Educ Res 9:020103

Redish EF (2003) Teaching physics with the physics suite. Wiley Inc., Somerset

Redish E (2005) Problem solving and the use of math in physics courses. Paper presented at world view on physics education of focusing on change, Delhi, pp 1–10

Redish EF, Kuo E (2015) Language of physics, language of math: disciplinary culture and dynamic epistemology. Sci Educ 24:561–590

Roberts AL, Sharma MD, Britton S, New PB (2007) An index to measure the ability of first year science students to transfer mathematics. Int J Math Educ Sci Technol 38(4):429–448

Shaffer PS, McDermott LC (2005) A research-based approach to improving student understanding of the vector nature of kinematical concepts. Am J Phys 73(10):921–931

Van Deventer J (2008) Comparing student performance on isomorphic math and physics vector representations. Electronic Theses and Dissertations 1348

Van Deventer J, Wittmann M (2007) Comparing student use of mathematical and physical vector representations. Paper presented at the physics education research conference, Greensboro, NC, vol 951, pp 208–211

Van Heuvelen A (1991) Learning to think like a physicist: a review of research-based instructional strategies. Am J Phys 59(10):891–897

Wemyss T, van Kampen P (2013) Categorization of first-year university students' interpretations of numerical linear distance-time graphs. Phys Rev ST Phys Educ Res 9:010107

Zavala G, Tejeda S, Barniol P, Beichner RJ (2017) Modifying the test of understanding graphs in kinematics. Phys Rev Phys Educ Res 13:020111

Research-Based Innovation in Introductory Physics Course for Biotechnology Students

Daniele Buongiorno and Marisa Michelini

Abstract Teaching introductory physics for biotechnology students requires to revise contents and methods in order to promote the developing of methodological competences through active participation of students. In the framework of the Model of Educational Reconstruction, such research-based innovation started three years ago in Udine University (IT). Structure, choices and research aspects related to the innovation will be here discussed.

1 Introduction

University teaching innovation is a need arising from the new missions characterizing a university accepting a huge number of students, the social and employment transformations produced by information and communication technologies (ICTs), the new ways in which students learn, the need to link secondary instruction and employment reality and the problem of the formative dropout. The innovation regards the structure, the resources and the organization that universities offer to student's participation as for example facilities, e-learning platforms, classrooms structure and transversal educational activities. Contents addressed in the single courses of study are very often isolated from the contexts of the specific course or the territory. Improving university teaching implies taking into account that student's personal involvement is an essential condition for an effective learning, the way in which a discipline is learnt is different from the way it has to be known, the information is no longer the focus of the teaching, but several resources are available, the learning is context-dependent and mostly that the goal is not to teach but make students able to learn: this cannot be simple achieved conveying information, but we have to build students' knowledge.

D. Buongiorno · M. Michelini (✉)
Physics Education Research Unit, DMIF (Department of Mathematics, Informatics and Physics), University of Udine, Via Delle Scienze 206, 33100 Udine, UD, Italy
e-mail: marisa.michelini@uniud.it

J. Guisasola and K. Zuza (eds.), *Research and Innovation in Physics Education: Two Sides of the Same Coin*, Challenges in Physics Education,
https://doi.org/10.1007/978-3-030-51182-1_14

169

In particular, improving university physics teaching is a need recognized at international level (http://hopenetwork.eu/content/hope-final-report) since low attention to educational aspects linked to teaching physics led the discipline to be taught in the same way (referring to its consolidated structure) in different courses of study, to privilege results with respect to processes, to use physical models in abstract contexts, with no links to the real world, to transmit formalization processes, approximations and simplifications rather than motivating them, making students experience the tools of the discipline. Physics is thus experienced as a discipline far from the real world, based on difficult laws that are not clear how and when to use.

A growing interest in Physics Education Research (PER) is focused on the role of physics as an introductory discipline in life-science areas as for example food or agricultural sciences, biology and biotechnology (Cummings et al. 2004; Redish and Hammer 2009; Watkins et al. 2012; Brewe et al. 2013; Donovan et al. 2013; Manthey and Brewe 2013; Meredith and Redish 2013; O'Shea et al. 2013; Thompson et al. 2013; Hoskinson et al. 2014; Redish et al. 2014; Michelini and Stefanel 2016). The need is not only to address the relevant topics for the specific field of study, in particular for those courses in which technological applications are prominent, but to build a competence to employ the methodological approach of physics to the biotechnological problems (Michelini and Stefanel 2018; Michelini and Stefanel 2016). A student-centered approach, in which the active role of students produces operative awareness of the role of physics in solving problems and relative instruments and methods, is relevant as well as taking into account the learning difficulties of students into the learning process to produce effective learning outcomes (Heron et al. 2004).

Teaching effectively introductory physics for biotechnology students is a multidimensional problem of "functional understanding" since it address the point of know how to correctly and coherently use physical concepts in the specific disciplinary-applicative context (McDermott and Shaffer 1992; McDermott et al. 2006). This task requires to (a) re-design physics education in order that its role can be recognized in the specific contexts characterizing the field of study with specific applications (Cummings et al. 2004; Meredith and Redish 2013; O'Shea et al. 2013; Hoskinson et al. 2014), (b) offer instruments and methods capable to build a physical competence in the sense of specific methodologies (Hoskinson et al. 2014) and (c) to point out strategies allowing an active role of students in learning with multitasking activities as for example the use of ICTs, laboratorial and problem-solving activities, ongoing evaluation and self-evaluation (Redish and Hammer 2009; Meredith and Redish 2013; Laws 2004).

To address these main needs, specific research-based intervention modules for innovation in introductory physics course for biotechnology in Udine University (IT) started in the academic year 2015/16. The content innovation pays attention to the strategies, and in particular to the role of problem-solving activities (Maloney 2011), the contribution of laboratorial activity (AAPT 1998), of ICT in teaching/learning process and the support of computer-assisted instruction (CAI) systems (Dunkel 1987). This work is on five different research plans: (a) curricular, understood as a global re-design of contents, instruments and methods; (b) learning modalities and strategies in teaching specific topics (i.e., fluid or optics); (c) role of the laboratorial activity; (d) link between academic research in biotechnology and (e) role of exercises. In particular, concerning exercises, they are a weakness of students but also a need to acquire operative competences.

The institutional role of innovation has been taken into account to give a guarantee of consolidation to the progress achieved in each phase of the experimentation.

2 General Design and Implementation

The global re-design of the introductory physics course for biotechnology students at University of Udine started in the academic year 2014–2015, when the formative success was about 30–40%, according to the theoretical framework of the Model of Educational Reconstruction (Duit et al. 2012). Based on PER results, research-based formative intervention modules (Anderson and Shattuck 2012) have been designed and implemented, taking into account the above-mentioned goals and aspects, focusing on increase the quality of teaching, avoiding reductionism promoting active involvement of students, both in presence, in laboratory and on the Web. Our goal is to build a functional understanding of the addressed contents, in order to allow students in gaining mastery in using physical concepts to critically interpret phenomena in their specific field of study, giving them the opportunity to use methodologies employed in physics in order to derive laws and experimentally validate them.

Our choices focused on instruments and methods (introducing an e-learning platform, individual and cooperative tasks), on the role of laboratory, which was given a wide importance according to (Cummings et al. 2004; Meredith and Redish 2013; Redish et al. 2014) and on the development of significant exercises to be proposed to students both via a CAI system and in flipped classroom.

The very first innovation in introductory physics course for biotechnology students occurred in the academic year 2015–16. At that time, the course consisted in 3 CTS (out of a total of 180 CTS during 3 years) for a total of 30 h. Results produced a recognition of role of physics and now, in the academic year 2017–18, introductory physics for biotechnology is a 4 CTS during the first semester of the first year. The course offers 52 h of activities, 26 of which are dedicated to laboratory activities, 20 to teaching and discussing, 6 to exercises and additional 14–20 to tutoring. Every year the sample consisted—on average—of 60 students, selected out of 200 applicants at the beginning of the academic year by means of a selection test with the same criteria at national level.

The role of the laboratory work increased during the three years introducing ICT-based experiments for studying light diffraction phenomena and absorption spectra and conduction of heat with real-time graphs. Experiments have been chosen in a three years long process in order to represent an interpretative challenge for students that do not have to follow a ready-made procedure but have to interpret and analyze data in order to produce a final report focusing on the specific formative elements.

The choice of focusing on the mastery of physical aspects involved in biotechnological contexts requires that the aspects concerning with typical experimental methods in physical analysis are experienced by students with a direct and active role. In this way, students do not only become aware of the experimental procedures, but they become competent in choosing inquiry methods, data analysis and re-elaboration procedures. Significant laboratorial activities have thus been chosen in which the work methodology was set according to ISLE method (Etkina and Van Heuvelen 2007) taking into account that, since they are freshmen, the cannot be completely empowered as concern the design phase. For this reason, the setting of the laboratorial part of the course consisted in an integration with the frontal lessons in which the basic physical principles founding each experiment were presented in general and concerning the specific experiment, only the goals, the available instruments and employment suggestions were provided to students, allowing them to point out the subtended problematic aspects, to choose the best data collection procedures, the conduction of the experiment and the relative reporting. Acting this way, the obvious puzzling of the students, deriving from the responsibility of managing and experiment and the relative data analysis, turned out in a cooperative learning involvement of the students groups, gaining mastery on the single disciplinary contents and on performing an experiment with relative data and results discussion.

Students were offered the opportunity to attend several seminars in order to link the academic teaching with advanced research fields in biotechnology as IR spectroscopy, PET, nanotechnologies and nano-structured magnetic materials providing individual written reports.

During the last academic year, a CAI system was designed and offered to students on an online platform. For every topic addressed during the course, three categories of exercises were available: multiple-choice questions for self-evaluation (thanks to an interactive interface showing students the right answers and giving them feedback), open exercises with numerical results and open exercises with discussion of the procedure (performed in class with a tutor). Exercises have a role of consolidation, deepening and self-evaluation.

In the academic year 2013–14, the initial curriculum covered an introduction on physics (units of measure, vectors, kinematic of the material point), forces and Newton's laws, linear momentum, kinetic and potential energy, systems of particles and rigid bodies (momentum of a force and conservation of the angular momentum), fluidostatic and fluidodynamics, electric field and potential, AC circuits with resistors and capacitors, magnetic field and magnetic force, chemical kinetics.

In the academic year 2014–15, several changes have been made: an initial schedule of the lessons was made available to students, new topics were introduced (optical and thermal phenomena), session of facultative tutoring for solving exercises were proposed to students, a seminar on IR spectroscopy was organized as well as the creation of a Web environment to deposit materials concerning the lessons. Three intermediate tests ((a) measures and kinematics, (b) energy, and optics, (c) optics, rigid body, electrostatic, circuits) and a relation on the seminar served to the ongoing evaluation of students.

The next academic year, i.e., 2015–16, new topics were included: oscillation and waves renouncing to the magnetism, the tutoring during the course was enhanced and two laboratorial activities were introduced to be performed in groups. A seminar on positron emission tomography was proposed to students as well as the one on IR spectroscopy. The three intermediate tests served to the ongoing evaluation ((a) measure and kinematics, (b) energy, oscillation and waves, (c) optics, rigid body, electrostatic, circuits).

In the academic year 2016–17, magnetism was re-introduced, and a new topic was inserted in the curriculum: optical spectroscopy, enhancing once more the time allocated for (facultative) tutoring. The number of laboratory experiments was raised to 7 (plus 3 to be performed at home), and students were offered the possibility to addend two seminars on "nano-structured surfaces and nano-fluids" and "magnetic phenomena and nano-particles for medical applications." Intermediate tests were reduced to two, but reports on laboratory contributed to the final evaluation. Final exam consisted in 6 questions concerning laboratory activities and 15 open-ended numerical exercises. To the final evaluation, both the final exam, the laboratory reports and reports on the attended seminar were considered.

The last academic year, i.e., 2017–18 saw the last version of the implementation: 19 h were devoted to frontal lessons, introducing AC circuits. 9 h were devoted to exercises and a CAI system allowed students to interactively train themselves with 16 exercises for every topic (divided into categories: (a) multiple choices, (b) open-ended with results and (c) resolved exercises). The e-learning platform allows students to train themselves with a selection of exercises taken from the research literature on physics education. Students can self-evaluate, thanks to the potentialities of the CAI system which reports the correct answer after having tried to resolve the exercises. Tutoring sessions were devoted to discuss the problems pointed out by students.

Laboratory activities covered a total of 30 h during which students performed 15 experiments (Table 1) out of which half of them are basic physical experiments, as the measure of volume, analysis of the transfer function (length-period) of a pendulum, refraction. The other half are characterizing and formative as the determination of the transfer function of an electronic instruments (calibration of thermometric probes), heat conduction in solids, spectroscopic and optical measure with online sensors (Buongiorno et al. 2018). Table 1 reports the proposed experiments during the last two academic years, witnessing the enhanced role of the laboratory in the whole course, maintaining a set core of experiments. The final exam allowed the individual evaluation of each students both on contents and on laboratory, since the exam consisted in 13 questions concerning laboratory activities and 15 open-ended numerical exercises. The questions concerning the laboratory activities were of four types: (a) obtain a law or a value from a set of sample data; (b) description of the physical principles of the experiment and/or the measure; (c) comparison between obtained values and expected ones, taking into account absolute and relative uncertainties; (d) comment on discrepancies between expected values and measured ones. To the final evaluation, at least 10 out of 15 group laboratory reports were considered.

Table 1 Comparison and description of proposed experiments

Experiment	Goal of the experiment
AC circuits: use of an oscilloscope and concept of electric impedance[2]	Build an AC-powered R-L-C series circuit. Evaluate theoretically the resonant frequency, measure the experimental one and compare the two values
Calibration of a flute[1,2]	Evaluate the transfer function of a conical container seen as an instruments of measure of volume: from a measure of length (apothem) it is possible to evaluate the contained volume of a fluid
Measure of volume[1,2]	Evaluate the volume of a spherical object in two ways (via geometrical measures and via immersion in a fluid) with absolute and relative uncertainties
Pendulum transfer function (length period)[1,2]	Determine the length of a pendulum wit 1 s period analyzing the transfer function between length and period of different pendulums
Calibration of a dynamometer (Hooke's law)[1,2]	Evaluating the elastic constant of a spring in different ways measuring displacement versus weight
Reflection[1,2]	Demonstrate the law of reflection using a mirror and several pins to represent different light rays
Refraction[1,2]	Obtain the index of refraction of a material using a half-moon shaped piece of plexiglass and several pins to represent different light rays
Motion on an inclined plane with friction[2]	Obtain the motion laws for an object falling along an inclined plane drawing on a strip of paper a series of equally time-spaced dots representing the motion. Evaluate the friction coefficient comparing the theoretical and experimental acceleration
Two-dimensional collision[2]	Analyze a collision between two spheres projecting on a plane the displacement vectors representing the linear momenta (Fig. 1)
Thermometric probes calibration[1,2]	Obtain the transfer function (digital value versus temperature) of digital thermometric probes (Fig. 2)
Heat conduction in solids[1,2]	Obtain the heat conduction coefficient in solid using thermometric probes in different places along a metallic bar subjected to a temperature gradient

(continued)

Table 1 (continued)

Experiment	Goal of the experiment
Online analysis of single-slit diffraction[1,2]	Determine the laws of diffraction using an online sensor measuring intensity of light versus position along a single-slit diffraction pattern (Fig. 3, left)
Analysis of LED spectra[2]	Measure the energy of light in the spectrum of different LEDs with a diffraction grating and correlate it with the threshold voltage
Digital analysis of absorption optical spectra[2]	Analyze the change in a continuous spectra filtered with colored filters using a digital spectrometer (Fig. 3, right)
Analysis of discrete spectra[1,2]	Measure the energy of lines in a gas-discharge lamps spectrum with an optical goniometer
Dynamical analysis of an oscillator[1]	Obtain the laws of motion and the parameters of spring-mass systems using digital position, velocity, acceleration and force sensors

Numbers in brackets indicates the academic year/s in which each experiment has been proposed: (1) for a.y. 2016–17 and (2) for a.y. 2017–18

3 Results

Concerning the academic year 2014–15, students not attending lessons were about 20%, while 18% of students did not reach the minimum score on the intermediate tests whose maximum participation was on the first one (79%) and the minimum one on the second one (66%). Only 30% of students did not pass the final exam with a formative success of 70% with an average score of 25.6/30 (Table 2).

Also, in the academic year 2015–16, 36% of students produced reports on the attended seminars. The formative success at the end of the course raised to 80% (see Table 3) and the decision to devote 1 CTS more to the course was taken (for a total of 4 CTS). Critical topics turned out to be measures, motion, waves and circuits.

Table 2 Data concerning final outcomes in academic year 2014–15

Students status	Test 1 (AVG score)	Test 2 (AVG score)	Test 3 (AVG score)	Seminar report (AVG score)	Verbalized (AVG score)
Attending	48(25.4)	40(21.6)	46(24.1)	46(25.2)	43(25.6)
Success	43	34	39	39	30
Success/attending (%)	90	85	85	85	**70**
Success/registered (%)	70	56	64	64	49

Number of registered students = 61

Table 3 Data concerning final outcomes in academic year 2015–16

Students status	Test 1	Test 2	Test 3	Laboratory reports	Recovery	Verbalized
Attending	47	49	44	46	9	54
Success	35	45	41	46	8	43
Success/attending (%)	74	92	93	100	89	**80**
Success/registered (%)	65	83	76	85	–	80

Number of registered students = 54

Table 4 Data concerning final outcomes in academic year 2016–17

Students status	Test 1	Test 2	Laboratory reports	Verbalized (AVG score)
Attending	58	53	61	56 (25.4)
Success	39	45	56	–
Success/attending (%)	67	85	92	**97**
Success/registered (%)	49	57	71	71

Number of registered students = 79

In the academic year 2016–17, the formative success was 97% with an average score of 25.4/30 (see Table 4). Critical topic turned out to be optical spectroscopy and DC circuits as concern the numerical exercises and heat conduction as concern laboratory procedures (Fig. 4).

Both in intermediate tests and in the final exam, the constant monitoring of the outcomes subdivided by categories of exercises, allowed to fine-tuning the teaching to a complete formation of the students. An aspect that emerged was linked to spectroscopy topics, in particular, to the formation of spectral lines, in which some aspects are better understood than others (Fig. 4).

The last academic year, i.e., 2017–18 saw a formative success of 84% with an average score of 26/30 (Table 5 and Fig. 7). Optical spectroscopy (in particular, the interpretation of spectra), fluidostatic and fluidodynamics are among the most problematic topics, according to Fig. 5. Concerning laboratory activities, the most controversial tasks, i.e., the ones in which students gained the lowest scores, were study of optical diffraction, the analysis of digital spectra, the calibration of a dynamometer and the calibration of thermal probes as shown in Fig. 6.

Every academic year there is a group of students ranging from 18 to 34% never attending the frontal lessons. With respect to students attending the lessons, it is noticed that when there is a high number of intermediate tests there is an evident auto-selection of students that attend the exams only when they feel well prepared, so there is a reduction of the number of students attending the last tests.

As concern the laboratory reports, results are always pretty satisfactory and the major difficulties emerge only when the interpretative role is more demanding as in the case of the experiments of the diffraction and of the digital spectrometer to analyze absorption spectra.

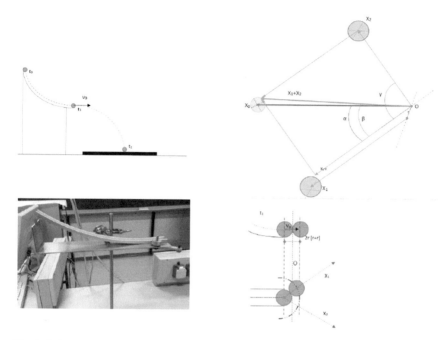

Fig. 1 Laboratory activity: analysis of a two-dimensional collision: the displacement vectors on the floor represent the linear momentum of the spheres. From a geometrical analysis, it is possible to determine the degree of elasticity of the collision and the percentage of lost kinetic energy

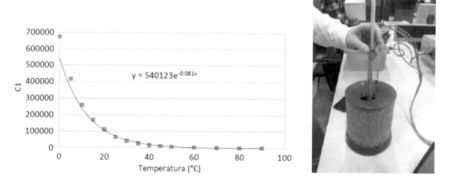

Fig. 2 Laboratory activity: calibration of a thermometric probe: the digital signal of the probe had to be correlated to the effective temperature, measured with a thermometer, in order to obtain the transfer function

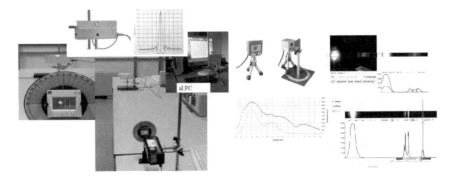

Fig. 3 Laboratory activities: optical diffraction (left) and optical spectroscopy (right). The employed devices are original equipment allowing a digital analysis of a diffraction pattern and of an optical spectrum (both continuous, discrete and a band spectrum, allowing also quantitative characterization of the transfer function of an optical filter)

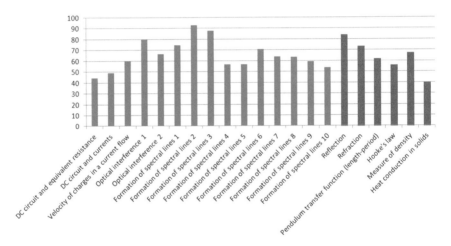

Fig. 4 Percentage of scores in the numerical exercises (blue) and in the laboratory exercises (red) in the final test of academic year 2016–17

4 Conclusions

The introductory physics course for biotechnology described here is not completely based on active learning, but there is the presence of numerous moments and situations in which the students, individually or in group, has the responsibility for the production of results and for operating decisions and making choices. The course turned from a traditional setting to a course centered on the role of physics in biotechnological areas and was focused on the active role of the students not only in the frontal lessons, but also in laboratory and at home, both individually and in group with reciprocal support, maintaining an individual evaluation of the theoretical and

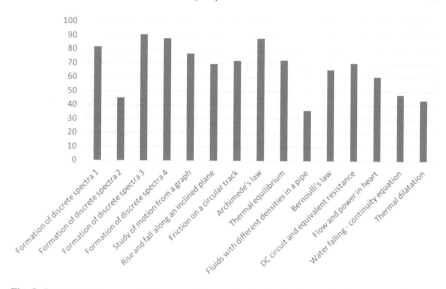

Fig. 5 Percentage of scores in the numerical exercises in the final test of academic year 2017–18

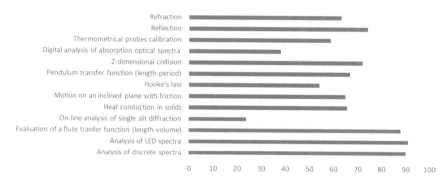

Fig. 6 Percentage of scores in the laboratory questions in the final test of academic year 2017–18

Table 5 Data concerning final outcomes in academic year 2017–18

Students status	Number of students
Registered	61
Attending	56
Pass final exam (failed)	47(9)
Success/attending	**84%**
Success/registered	78%

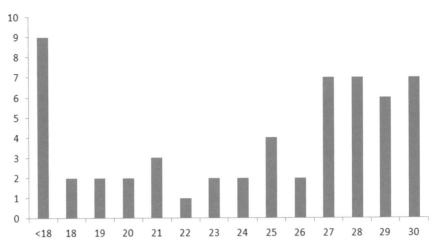

Fig. 7 Distribution of scores in the final test of academic year 2017–18

experimental competences. The role of laboratory turned out to be very useful to the extent that methodologies and data are analyzed critically.

Written problems for final exams are taken in the international literature to have a standard as a reference. Seminars turned out to be useful, since students were stimulated in searching for applications and reporting them. Exercises proposed to students allow them to work independently showing a high level of commitment asking for doubts, problems and questions. Exercises turned out to be very useful, and for the future is desirable to set out a system of credits using the possibilities given from the CAI system. Problem-solving activities, the CAI system and flipped classroom can turn out to be a good ensemble to manage exercises during tutoring activities and/or at home, even if students tend to prefer to follow a guided procedure for solving exercises rather than get involved on their own.

As concern contents, some of them can be left out as motion, dynamics and universal gravitational, since they are well addressed in secondary school, while other topics have to be deepen as ideal and real fluids, fluidodynamics, physical optics and spectroscopy, thermal phenomena and heat conduction.

Another important aspect is represented by the various seminars concerning biotechnological applications of physical concepts that students were able to attend, with the specific request of producing a report focusing on the underlying physics.

The innovation resulted in an increased formative success from 62 to 84% as concern for the last academic year.

Research carried out offers hints on the following aspects: (1) How to select and address the topics and in particular by the test-in, it emerged that it was possible to lighten mechanics and dynamics topics, but treating it in different ways since some open knots persisted; (2) how laboratorial activities offer a methodological contribution to the described goals; (3) how to engage students in problem-solving activities for every specific content; (4) how do students make use of the offered opportunities (seminars, encounters with researchers) to elaborate and deepen the studied topics; (5) how to design a structure for an introductory physics course for biotechnology students in terms of contents, instruments and methods.

Acknowledgements The authors are particularly grateful to the coordinator of the degree course in Biotechnology of University of Udine Gianluca Tell for allowing this innovation.

References

AAPT (1998) Am J Phys 66:483

Anderson T, Shattuck J (2012) Edu Res 41(1):16

Brewe E, Pelaez NJ, Cooke TJ (2013) CBE Life Sci Edu 12(2):117

Buongiorno D, Gervasio M, Michelini M (2018) SPETTROGRAFO: a prototype for a digital spectroscope in pubbl. In: Proceedings of FFP15, Orihuela, Spain

Cummings K, Laws PW, Redish EF, Cooney PJ, Taylor EF (2004) Understanding physics. Weley, Hoboken, NJ, USA

Donovan DA, Atkins LJ, Salter LY, Gallagher DJ, Kratz RF, Rousseauu JV, Nelson GD (2013) CBE Life Sci Edu 12(2):215

Duit R, Gropengießer H, Kattmann U, Komorek M, Parchmann I (2012) In: Jorde D, Dillon J (eds) Science education research and practice in Europe. Sense Publishers, Rotterdam, pp 13–37

Dunkel P (1987) Mod Lang J 71(3):250

Etkina E, Van Heuvelen A (2007) Getting Started in PER, Reviews in PER 1(1)

Heron PRL, Shaffer PS, McDermott LC (2004) Building excellence in undergraduate STEM education. AAAS, Washington, pp 33–38

Hoskinson AM, Couch BA, Zwickl BM, Hinko K, Caballero MD (2014) Am J Phys 82(34):434

http://hopenetwork.eu/content/hope-final-report

Laws PW (2004) Undergraduate science, technology, engineering and mathematics education. AAAS, Washington, pp 247–252

Maloney D (2011) Getting started in PER, Reviews in PER 2(1)

Manthey S, Brewe E (2013) CBE Life Sci Edu 12(2):206

McDermott LC, Shaffer PS (1992) Am J Phys 60(11):994

McDermott LC, Heron PRL, Shaffer PS, Stetzer MR (2006) Am J Phys 74(9):763

Meredith DC, Redish EF (2013) Phys Today 66(7):28

Michelini M, Stefanel A (2016) In: Dębowska E, Greczyło T (eds) Key competences in physics teaching and learning. University of Wrocław, Wrocław, Poland, pp 142–149

Michelini M, Stefanel A (2016) Proceedings of XXX Convegno Didamatica, Udine

Michelini M, Stefanel A (2018) Proceedings of II WCPE, Sao Paolo

O'Shea B, Terry L, Benenson W (2013) CBE Life Sci Edu 12(2):230

Redish EF, Hammer D (2009) Am J Phys 77(7):629

Redish EF et al (2014a) Am J Phys 82(368):368

Redish EF et al (2014b) Am J Phys 82(5):368

Thompson KV, Chmielewski J, Gaines MS, Hrycyna CA, LaCourse WR (2013) CBE Life Sci Edu 12(2):162

Watkins J, Coffey JE, Redish EF, Cooke TJ (2012) Phys Rev ST Phys Edu Res 8(1):010112

Investigating the Interplay of Practical Work and Visual Representations on Students' Misconceptions: The Case of Seasons

Silvia Galano, Francesca Monti, Giacomo Bozzo, Claudia Daffara, and Italo Testa

Abstract The paper aims at investigating to which extent the interplay between practical work activities and visual representations is effective in addressing with students' misconceptions about seasonal changes. To this aim, we designed a 10 h teaching sequence in which the students: (i) explore the radiation flow using a photovoltaic panel as a function of the incidence angle and distance from the source and (ii) use specially designed visual representations about seasonal change. A sample of about 88 prospective primary teachers was involved in this study. Analysis shows a significant impact of the practical activities but almost no effect of the different kind of the used images. Implications for future research are briefly discussed.

1 Introduction

Previous research studies have shown how much students' misconceptions about seasonal changes are resistant to traditional teaching (Atwood and Atwood 1996). Moreover, some authors have pointed out the difficulty in reading and interpreting images as a relevant factor that may influence the persistence of such misconceptions (Ojala 1992). In two previous studies (Testa et al. 2015; Galano et al. 2018), we showed that: (i) practical activities may enhance students' understanding of the mechanism underlying seasonal changes and (ii) specially designed images may be more effective than textbook images in helping students overcome typical misconceptions about seasons, as that in summer the Earth is closer to the Sun. In this paper, we aim at investigating the effectiveness of a teaching sequence that blends both

S. Galano · I. Testa (✉)
Department of Physics, "E. Pancini" Federico II University of Naples, Complesso Monte S. Angelo, via Cintia, 80126 Naples, Italy
e-mail: italo.testa@unina.it

F. Monti · G. Bozzo · C. Daffara
Computer Science Department, University of Verona, Verona, Italy

183

J. Guisasola and K. Zuza (eds.), *Research and Innovation in Physics Education: Two Sides of the Same Coin*, Challenges in Physics Education,
https://doi.org/10.1007/978-3-030-51182-1_15

practical activities and specially designed images. The research question that guided the study was:

- *To what extent the interplay between practical work activities and innovative visual representations is effective in addressing students' misconceptions about seasonal changes?*

2 Methods

In this paragraph, we present the instructional context and the data analysis.

2.1 *Instructional Context*

For this study, we designed three instructional contexts: (1) practical work + textbook images; (2) practical work + specially designed images; and (3) practical work + no images. A brief booklet was distributed to summarize the explanation of the seasonal changes.

2.1.1 Practical Work Activity

During this activity (Testa et al. 2015), the students first investigate the dependence of the flow of a radiation hitting a surface at a fixed distance from the source, on the inclination between the normal to the incidence surface and the direction of the radiation. Then, they investigate how the flow varies when the surface is kept orthogonal to the direction of the incident radiation, and the distance between the source and the surface changes. The same source is used in both experiments. The aim is to compare the two laws:

$$\frac{P(\theta)}{P_0} = \cos(\theta) \tag{1}$$

$$P(D) = \frac{A}{D^2} \tag{2}$$

where P is the power received by the surface; θ is the angle between the normal to the surface and the direction of the incident radiation; D is the distance between the source of the radiation and the surface, P_0 and A are constants that depend on the power emitted by the source and the geometry of the sensible area of the surface, respectively.

The laws in (1) and (2) are empirically derived by the students. In particular, a photovoltaic panel is used as surface, while an incandescent light bulb is used as radiation source. The bulb represents the laboratory "Sun", while the panel represents a small area of the Earth's ground. The experimental setup is reported in Fig. 1, while the plot of example experimental data obtained by different groups of students when investigating the dependence of the absorbed power on the panel inclination is reported in Fig. 2.

Then, using Eq. 1, the students are asked to evaluate the normalized difference

$$1 - \frac{P(\theta_w)}{P(\theta_s)} = 1 - \frac{\cos(\theta_w)}{\cos(\theta_s)} \tag{3}$$

Fig. 1 The experimental setup used during the practical activities

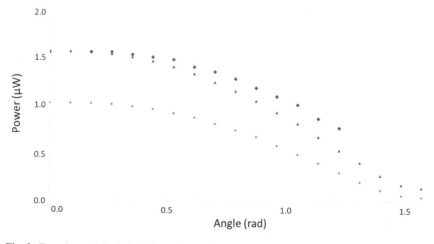

Fig. 2 Experimental data of the dependence of the power incident on the panel when varying the incidence angle

for five locations of the Earth (tropics, Equator, Arctic, and Antarctic Circle) at two specific times of the year, defined by the angles θ_w and θ_s. The chosen times are: winter and summer solstices for tropics and Equator, summer/winter solstices and autumn/spring equinoxes for Arctic and Antarctic Circle, respectively. For instance, for the cancer tropic we have:

$$\theta_w = 46° \rightarrow \cos(\theta_w) \cong 0.69$$
$$\theta_s = 0° \rightarrow \cos(\theta_s) \cong 1 \tag{4}$$

so that Eq. 3 gives a normalized difference of about 30%.

Then, using Eq. 2, the students are asked to calculate the normalized difference:

$$1 - \frac{P\left(D_{\text{Aphelion}}\right)}{P\left(D_{\text{Perihelion}}\right)} = 1 - \left(\frac{D_{\text{Perihelion}}}{D_{\text{Aphelion}}}\right)^2 \tag{5}$$

Using the values $D_{\text{Aphelion}} \cong 1.02$ u.a. and $D_{\text{Perihelion}} \cong 0.98$ u.a., Eq. (5) gives a normalized difference of about 7.6% which is significantly lower than that obtained from the variation due to axis' tilt. In such a way, the distance misconception can be quantitatively addressed.

2.1.2 Specially Designed Images

Two commonly used textbook images about seasonal changes were chosen for context (1). For context (2), we designed two diagrammatic images (Figs. 3 and 4).

The image in Fig. 3 has many differences with respect to usual textbook images. First, it features an Earth's orbit which is circular and not elliptical. Moreover, it does not include arrows to indicate the rotation and revolution of the Earth and other information not relevant for the phenomenon (e.g., segments that connect the Earth to the Sun). It also avoids the use of text as, for instance, the reference to aphelion and perihelion. The adopted perspective emphasizes the constant direction of the axis during the motion. To prevent misleading ideas about a wrong axis' tilt, the image presents a different viewpoint at the bottom of the image: in particular, we chose the viewpoint of an observer on the same plane of Earth' orbit. In Fig. 4, the different incidence of sunrays due to axis' tilt is shown by explicitly reporting the angle between the direction of sunrays and the plane tangent to the Earth surface at two different times of the year, winter and summer solstices, to highlight the largest difference in illumination at that particular place. To help students connect the inclination of sunrays along the Earth surface and their own experience with different incidence of Sun radiation on the ground, the corresponding view of an observer on the Earth is reported at the bottom of the image. Using both Figs. 3 and 4, the role of the two factors—orbit and axis' tilt—on seasonal change may be reinforced.

Fig. 3 Image designed to explain the change of season. Translation of Italian labels: Sun (Sole), Earth (Terra)

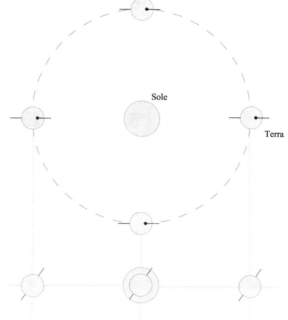

Fig. 4 Image designed to explain variable incidence of Sun's radiation on Earth during the year. Translation of Italian labels: December 21st Latitude 40° N (bottom left panel), June 21st Latitude 40° N (bottom right panel)

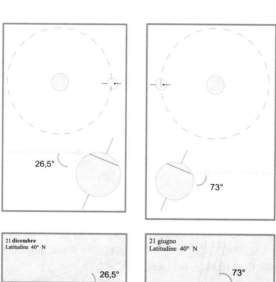

Table 1 Rubric adopted to analyze students' explanations of seasonal changes

Category	Example
Unclear	Because the Earth rotates
Incorrect	The Earth is closer to the Sun during summer
	During winter it is colder, since the Earth does not face the Sun
Partial	Sunrays are more inclined in winter than in the summer
Correct	Because the sunrays inclination changes due to axis' tilt
	Sunrays are inclined due to revolution and axis' tilt

2.2 Instrument and Data Analysis

To answer our research question, we adopted a pre-post design using an instrument that featured three tasks: a written task, a drawing task, a questionnaire featuring multiple-choice and true/false items. In the written task, students were asked to explain the change of seasons in words. In the drawing task, students were asked to explain the phenomenon using a drawing. Two multiple-choice and six true/false items completed the instrument. To analyze students' drawings and explanations we used the rubrics reported in Tables 1 and 2 (Galano et al. 2018). For these tasks, inter-rater reliability was evaluated through Cohen's kappa. Resulting values were 0.7 for drawings and 0.8 for explanations. The scoring of the responses to the questionnaire was as follows: 0 point for an incorrect answer, 1 point for each correct response to a true/false item and for distance-based answer choice in the multiple-choice items; 2 point for a correct answer choice in the multiple-choice items. The maximum score was 10.

2.3 Sample

A convenience group of 88 prospective primary teachers, attending a Foundations and Didactics of Physics course, was involved in the study. Average age was 20. The participants were randomly assigned to the three teaching contexts: practical work + textbook images, $N = 30$; practical work + specially designed images, $N = 30$; practical work + no images, $N = 28$.

3 Results

In this paragraph, we present the results of the study.

Table 2 Rubric adopted to analyze students' drawings of seasonal changes

Category name	Drawing	Description
Distance		i. Earth and Sun are represented only once at different relative distance according to the season (closer → summer; farther → winter). Symbols representing rotation or revolution may be also present. The main underlying idea seems that seasons are due to the relative positions of the Sun and the Earth ii. More than one Earth orbiting around the Sun are represented. Segments representing variable distance between Earth and Sun may be present. The orbit is represented as elliptical. Verbal elements that indicate the different seasons may be present. The main underlying idea seems that seasons are due to the relative positions of the Sun and the Earth
Inclination		i. Only inclined or convergent sunrays on Earth surface are represented. The main underlying ideas seems that change of seasons is related to variable inclinations of sunrays during the year ii. Only the Earth is represented with its inclined axis. Indication of poles and of the tilt is also present. Segments representing sunrays are also present. The main underlying idea seems that seasonal changes are due to the axis' tilt

(continued)

Table 2 (continued)

Category name	Drawing	Description
Correct	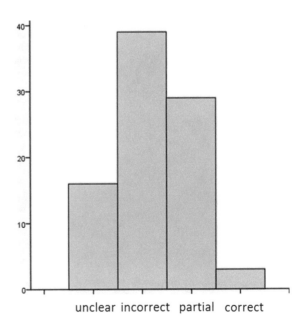	i. One or more orbiting Earths around the Sun are represented with indication of the tilt of Earth's axis. The orbit is represented as elliptical. Verbal elements that indicate the different seasons are present. Symbols indicating rotation or revolution are present. The main underlying idea seems that seasons are due both to Earth's orbit and axis' tilt ii. More than one Earth orbiting around the Sun are represented. The orbit is represented as elliptical. Verbal elements that indicate the different seasons are present. The main underlying idea seems that seasons are due to Earth's orbit

3.1 Written Task

Figures 5 and 6 report the distribution of students' answers to the pre-test and post-test written task for the whole sample. The results show a significant improvement (marginal homogeneity test, MH = 6.452, $p < 10^{-4}$). Figures 7 and 8 report the

Fig. 5 Pre-test categorization of students' explanations

Fig. 6 Post-test categorization of students' explanations

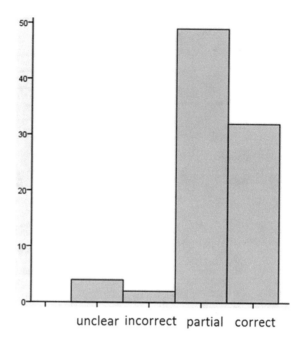

Fig. 7 Pre-test distribution of explanations for the three groups of students

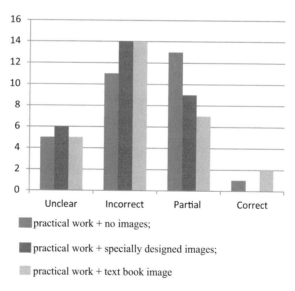

distribution of students' explanation across the three groups. The differences are not significant for both pre-test and post-test ($\chi^2 < 10.124$).

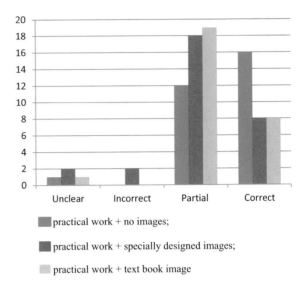

Fig. 8 Post-test distribution of explanations for the three groups of students

- practical work + no images;
- practical work + specially designed images;
- practical work + text book image

3.2 Drawing Task

Figures 9 and 10 report the distribution of answers to the pre-test and post-test drawing task for the whole sample. The results show also for this task a significant improvement before and after the instruction (marginal homogeneity test, MH = 7.056, $p < 10^{-4}$). Figures 11 and 12 report the distribution of students' drawings

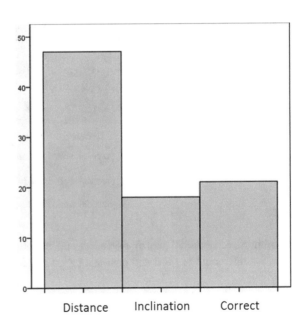

Fig. 9 Pre-test categorization of students' drawings

Fig. 10 Post-test
categorization of students'
drawings

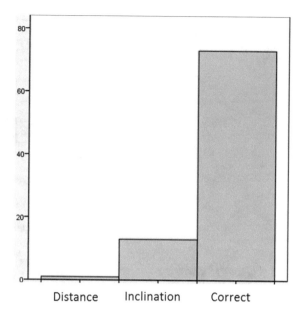

Fig. 11 Pre-test
categorization of drawings
for the three groups of
students

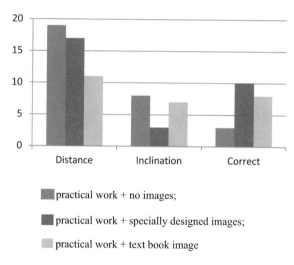

■ practical work + no images;

■ practical work + specially designed images;

▢ practical work + text book image

across the three groups. The differences are not significant for both pre-test and
post-test ($\chi^2 < 7.751$) also for this task.

Fig. 12 Post-test categorization of drawings for the three groups of students

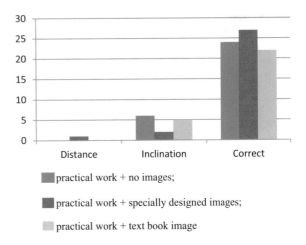

☐ practical work + no images;

☐ practical work + specially designed images;

☐ practical work + text book image

3.3 Questionnaire

Average pre-test score for the whole sample was 7.8 ± 1.2 (st. dev), while the post-test score was 9.6 ± 0.8 (st. dev). The difference, as for the other tasks, is statistically significant ($t = 12.218, df = 84, p < 10^{-4}$). However, the differences across the three groups are not statistically significant in both pre-test and post-test ($F(2) < 4.735, p < 0.233$).

4 Discussion and Implications

Overall, collected evidence supports the conclusion that the proposed practical activities were effective in improving students' knowledge about seasons for the three groups. In particular, the distance misconception seems to have been addressed, since its frequency decreased between pre-test and post-test from about 50% to less than 5%. Such evidence confirms the results of our first previous study (Testa et al. 2015), in particular it supports the effectiveness of the proposed measurements with the solar panel and the bulb. However, there were no effects detected between the three experimental groups. This result contradicts the results of our second previous study (Galano et al. 2018), in which we found significant differences in students' performances when exposed to different support in using the images. Such evidence suggests that the proposed laboratory activity is predominant with respect to the support provided by the specially designed images. This result has at least two interpretations. First, as suggested by literature (Bowker 5), our findings confirm that the influence of conventional representations on students' drawings seems to decrease as instructional support increases (Bowker 5). Second, in the study described in (Galano et al. 2018), the involved students were 13–14 years old, while the students involved in this study were older (see methods section) and had been exposed to high school teaching. As such, the sample in this study may have developed higher cognitive skills that helped them better interpret the textbook images or the booklet, which did not feature any supporting image.

A first implication of the findings of the present study concerns the teaching of familiar astronomical phenomena. In particular, our evidence suggests that practical activities are effective to address well-known misconceptions in astronomy—as the distance misconception for seasonal changes. The proposed activity with the solar panel can also be easily adapted to teaching contexts in which it is not possible to introduce the proposed formalization, for instance, by integrating it with the use of suitable computer-based simulations. Second, the collected evidence warrants more research about the interplay between practical activities and the use of images. We are currently planning a replication of this study with middle school students (13–14 years old) to see whether the role of the practical activity is predominant also for this population of students. Finally, the study could be fruitfully extended to other astronomy phenomena, e.g., Moon phases, for which literature suggests a relevant influence of textbook images on students' conceptions.

References

Atwood RK, Atwood VA (1996) J Res Sc Teach 33:553

Bowker R (2007) Env Ed Res 13:75

Galano S, Colantonio A, Leccia S, Marzoli I, Puddu E, Testa I (2018) Phys Rev Phys Ed Res 14:010145

Ojala J (1992) Int J Sc Ed 14:191

Testa I, Busarello G, Puddu E, Leccia S, Merluzzi P, Colantonio A, Moretti MI, Galano S, Zappia A (2015) Phys Ed 50:179

Using Theory to Inform Practice in the Advanced Physics Classroom

Ramon E. Lopez, Michael A. Greene, and Ximena Cid

Abstract Physics education research has focused much more on lower-level, introductory courses as compared to upper division and graduate physics education. However, there are general principles and findings that extend across all areas of learning, such as the strong evidence in favor of active learning environments. But taking the theoretical basis and pedagogical strategies generated by research at one level of education and applying it to create a learning environment appropriate to upper division and graduate physics courses requires careful consideration of the issues facing students, and the instructor, in such courses. For example, the motivations of students in an introductory course are very different from the motivations of students in a graduate course. The number of students in a classroom is often quite different. The size of the research base on student difficulties and the amount of research-based instructional resources available to an instructor will be different. In this paper, we discuss several examples of the application of research-based techniques to classroom instruction in upper division and graduate physics courses, how the specifics of the student audience have resulted in modifications of the pedagogical approach, and the student response to these instructional strategies.

1 Introduction

The study of mechanics, force, and motion is the starting point in physics education, and it is not unreasonable to say that the study of student difficulties in acquiring a Newtonian understanding of force and motion was the starting point for physics education research (Halloun and Hestenes 1985). It is well known that

R. E. Lopez (✉) · M. A. Greene
Department of Physics, University of Texas at Arlington, Arlington, TX 76019, USA
e-mail: relopez@uta.edu

X. Cid
Department of Physics, California State University Dominguez Hills, Carson, CA 90747, USA

J. Guisasola and K. Zuza (eds.), *Research and Innovation in Physics Education: Two Sides of the Same Coin*, Challenges in Physics Education,
https://doi.org/10.1007/978-3-030-51182-1_16

197

students learning physics have difficulty with creating coherent conceptual frameworks (McDermott 1993). Over the past few decades, a number of research-based, active learning instructional approaches have been developed to deal with these difficulties (McDermott and Shaffer 1992; Crouch and Mazur 2001; Perkins et al. 2006; Otero et al. 2010). The research in physics education has concentrated primarily on the introductory lower division courses, though there are examples of work done in upper division physics topics (Ambrose 2004; Singh et al. 2006). However, the great majority of students taking physics courses are in lower division courses. Thus the techniques and materials developed that have been widely disseminated (Finkelstein and Pollock 2005) have focused on students in those courses, along with most of the work on the effectiveness of research-based pedagogical innovation (Deslauriers et al. 2011), even though active learning methods can work quite well with more advanced students (Lopez and Gross 2008).

One such recent innovation is called the "flipped" classroom (Tucker 2012). In a flipped classroom the passive, lecture portion of the course is done outside of class, generally with student viewing pre-recorded lectures. The time in class is reserved for problem-solving that in a regular course would constitute the "homework" portion of the course structure. Thus, the class is "flipped" relative to a traditional structure. For the most part, the flipped classroom approach and research into its effectiveness has focused on high school or lower division undergraduate courses (Bishop and Verleger 2013), with the exception of graduate medical education (Tune et al. 2013), which tends to adopt a wide range of innovative educational practices (West et al. 2000).

In this paper, we report upper division undergraduate and graduate physics students' response to both active engagement techniques and to a flipped classroom structure. We first describe an experiment in which graduate students engaged in sample class taught in an active engagement style. The student response to this experiment led to the modification of the graduate Classical Mechanics class to change it from a traditional course to an active learning environment. The course was then flipped, with lectures posted online and classroom time reserved for clarification of lecture material and collaborative problem-solving. After a couple of years, the upper division undergraduate course was also flipped. Students who took these courses were surveyed, and it is the results of that survey that we will report.

2 Graduate Student Response to a Sample Active Learning Class

In the Fall of 2010, an experiment was conducted with graduate students in the UT Arlington physics Ph.D. program. We wanted to gauge the student response to an active learning environment on a topic appropriate for graduate instruction. Seven students volunteered to take part, 3 US students and 4 international students. All of them had completed their required coursework, had passed the departmental qualifying exam (to be described in more detail later), and were working on their dissertation research. Two of the seven had had an undergraduate class that used non-traditional instruction (something beyond traditional lecture), but the other five had had only traditional lecture courses in their education. None of these students had ever studied plasma physics, which was the subject matter for the sample class.

The sample class focused on one-fluid, ideal magnetohydrodynamics (MHD), and the derivation of the energy equation in conservative form. This derivation is a straightforward exercise provides a well-defined "chunk" that was amenable to our experiment. We began with the MHD momentum equation for an isotropic plasma:

$$\rho \frac{\mathrm{d}\vec{V}}{\mathrm{d}t} = -\nabla P + \vec{J} \times \vec{B} \tag{1}$$

where ρ is the mass density, $\vec{V}$ is the plasma bulk flow velocity, P is the isotropic (scalar) plasma pressure, $\vec{J}$ is the current per unit volume, and $\vec{B}$ is the magnetic field. Equation (1) is essentially $\vec{F} = m\vec{a}$ per unit volume, where the forces are the pressure gradient force and the magnetic force on a current, which was discussed with students at the outset. If one takes the dot product of the momentum equation with the plasma velocity, using the adiabatic equation of state, $P\rho^{-\gamma} = \text{constant}$, and the Maxwell equations, one can arrive at the following expression:

$$\frac{\partial}{\partial t}\left(\frac{\rho V^2}{2} + \frac{P}{\gamma - 1} + \frac{B^2}{2\mu_o}\right) + \nabla \cdot \left(\frac{\rho V^2}{2}\vec{V} + \frac{\gamma P}{\gamma - 1}\vec{V} + \frac{\vec{E} \times \vec{B}}{\mu_o}\right) = 0 \tag{2}$$

The way in which this expression was derived in the sample class was not through the traditional lecture mode. At each step, the instructor (R. Lopez) had the students dialog in small groups as to what the next step in evaluating each term might be once we had taken the dot product of the momentum equation with the velocity. However, rather than open-ended "what do you think" or "who has the next step" kind of probing, a number of multiple-choice questions were posed to students along the lines of peer-instruction questions (Mazur 1997). One example concerns the term $\vec{V} \cdot \nabla P$, which is part of the total time derivative of the pressure. That is where the adiabatic equation of state comes in, since the total time derivative of $P\rho^-$ is zero. Several answers were provided from which the students had to pick one, with common errors in differentiation used as distractors.

At each step, students were asked to discuss in the small groups (either 3, 2, 2 or 3, 4 students—it varied through the process) the meaning of things like "adiabatic equation of state." What does it mean? Some students remembered that it had to do with heat transfer, but were not certain. We discussed that that equation of state arises when there is no energy transfer into or out of the ideal gas in question. At the end, we had Eq. (2) in conservative form (a partial time derivative of densities plus the divergence of fluxes adding to zero). The students were told to pick a term in the equation and in their small groups provide a simple, physical description of what the term means. Through this process, which took several back and forth iterations, the students arrived at a physical description of each term. The students came to the realization that the terms in the partial time derivatives are the kinetic energy density, thermal energy density, and (electro)magnetic energy density, respectively. The terms inside the divergence are the fluxes of these quantities. Thus, Eq. (2) is an example of "what goes in, goes out, except what stays there" and is just a way for writing the conservation of energy that stems from the adiabatic equation of state.

The entire exercise took 80 min. To determine student reaction to the sample class, over the next few days each student was interviewed on videotape by X. Cid (with anonymity preserved). The interviews were then transcribed. All of the students reacted positively and felt that they were engaged during the exercise. Some typical comments were as follows:

> So he was relating the physical world to the math world and giving some explanation or some connection between the two…. So that is what I expect from a physics class…the math and the physics and the two of them at the same time.

> While I was in the class, I had to concentrate on the topic that he was teaching and I had to listen to the opinion given by the other group members…Yeah so it was more engaged…the best thing that I found in that class was it was engaging and that is the most important part of the learning.

Another student reported the following:

> Before I had this lecture, I mean of course we have talked about conservation…but I never put…I actually never put the two together…it seemed like I made some connections that I've never made before…just from the hour…half hour lecture…

This comment is quite remarkable. Conservation of energy is a topic that appears throughout the physics curriculum, and we assume that physics graduate students have a solid grasp of this essential, elementary principle. But here we have an advanced graduate student, not far from completing a Ph.D., after having passed all the required classes and exams, stating that a single lecture has led to making "… some connections that I've never made before …" regarding conservation of energy. Evidence gained from working with graduate students as teaching assistants has demonstrated that many physics graduates still have incomplete or even mistaken ideas about phenomena and concepts (McDermott 2001), indicating that graduate students could benefit from more active, rather than passive, classroom environments, just as undergraduates do.

3 The Introduction of the Flipped Model and Collection of Student Response Data

In the Fall 2013 semester, the graduate Classical Mechanics course was converted to an active lecture format using the same approach as described in the sample class, including some time spent on collaborative problem-solving and review of homework. The book used (Goldstein, Poole, and Safko, *Classical Mechanics*, 3rd

Edition) is the standard book used in the course. After one semester experimenting with active learning in the classroom in the form of collaborative problem-solving and peer-instruction sequences, the course was converted to a flipped model for the Fall of 2014 (the course is taught in the fall, once per year), in which the lectures were put online and the class time was used to work problems and clarify questions arising from the lectures, but no formal lectures were presented. In the Spring 2016, semester the same approach was implemented in the undergraduate advanced mechanics course (Marion and Thornton, *Classical Dynamics*, 5th Edition). That course is taught each spring, and both courses were taught by the same instructor (R. Lopez).

The original undergraduate course had included many aspects of active learning, such as some collaborative problem-solving during class. The lecture portion included an active component as well, with questions using the peer-instruction technique which had been developed over several years to address student difficulties with advanced topics like Lagrangians and Hamiltonians (Lopez 2008). These concept questions aim either at probing conceptual understanding or the issue of transfer of mathematical procedural knowledge to the physics context. The transfer problem is a big issue for students because if they cannot understand how the math relates to the physics, they cannot succeed in building the required frameworks around the new material. Often the students are not quite sure how to manipulate functions like Lagrangians, and simple, multiple-choice questions discussed in class can clarify the situation. An example of such questions is in Fig. 1, where students are to determine the canonical momentum associated with the θ coordinate for a pendulum.

The online lectures for the undergraduate flipped class contained these questions with instructions to pause the video and write down an answer to the question before proceeding to the next portion in which the answer was discussed. Similar attempts were made to incorporate active lecture techniques into the graduate flipped lectures, but there was less material to work (basically what had been developed in one semester) with so there were not as many embedded questions/activities in the graduate lectures. However, one of the negative aspects of migrating lectures to an online format is that the small group discussions that occur in class around the peer-instruction concept questions are lost since students view the lectures individually.

4   Findings from the Survey

After the end of the Spring 2016 semester, once grades had been submitted, a survey was given to both graduate and undergraduate students to collect data on their response to the flipped mode of instruction. The total enrollment in the two semesters (Fall 2014 and Fall 2015) of the flipped graduate course was 22, and the enrollment in the flipped undergraduate course (Spring 2016) was 20. A total of 24 students filled out the survey (12 undergraduate and 12 graduate students), providing a response rate of just over 50%.

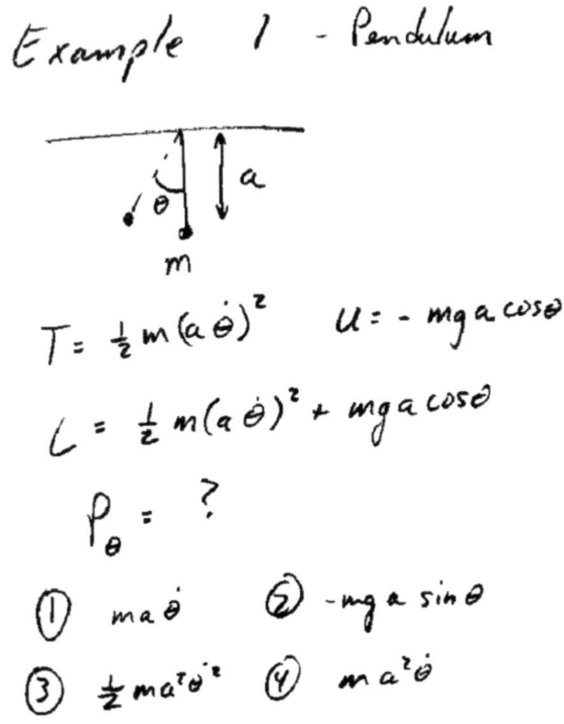

Fig. 1 Lecture question for
undergraduate advanced
mechanics

$Example \quad 1 \quad - \quad Pendulum$

$$T = \tfrac{1}{2} m (a\dot{\theta})^2 \qquad U = - mga\cos\theta$$

$$L = \tfrac{1}{2} m(a\dot{\theta})^2 + mga\cos\theta$$

$$P_\theta = ?$$

① $ma\dot{\theta}$ ② $-mga\sin\theta$

③ $\tfrac{1}{2} ma^2\dot{\theta}^2$ ④ $ma^2\dot{\theta}$

The survey comprises 22 statements which students ranked on a five-point Likert scale, with $1 =$ Strongly Disagree and $5 =$ Strongly Agree. Statements were organized in four categories: affective (e.g., I enjoyed this flipped classroom), participatory (e.g., I watched every video before class), cognitive (e.g., solving problems in class was helpful for my understanding of the topics), and procedural (e.g., while watching the videos, I took notes on paper). The student responses were tabulated, and responses were compared to the null hypothesis, a neutral response (ave $= 3$), to determine if the students as a group agreed or disagreed with a given statement. In the findings below we report the average response and the p-value of the comparison (t-test) to the null response indicating if a difference between the null value and the average student response is statistically significant.

In response to the statement "I enjoyed this flipped classroom" (ave $= 2.91, p = 0.747$), the students either liked or disliked the flipped classroom relative to traditional classrooms, but there was no overall preference. However, the spread in responses ranged from 1 to 5, with a standard deviation of 1.28. Thus some students really liked the flipped classroom, but an equal number really disliked it. Moreover, while the response to the statement "I learned more from the flipped classroom compared to what I learn from a traditionally taught lecture class" (ave $= 2.70, p = 0.178$) also showed no overall preference relative to the neutral response, it was correlated with whether they enjoyed the flipped class model or not, as can be seen in Fig. 2.

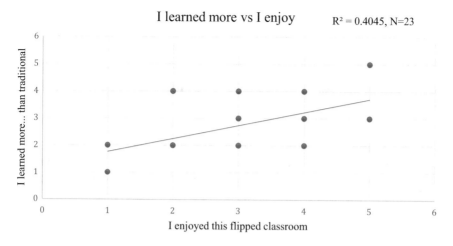

Fig. 2 Correlation between reporting learning versus enjoyment of the flipped model

The students did do what was asked of them, and they broadly agreed with the statements that we classified as participatory. Students viewed the videos regularly and did not "binge watch." They agreed with "I watch every lecture before class" (ave $= 4.26$, $p = 0.000$) and disagreed with "I watched the lecture videos only on the weekends" (ave $= 2.04$, $p = 0.000$). Students actively took notes (ave $= 4.52$, $p = 0.000$), paused the video to think and answer concept questions (ave $= 4.48$, $p = 0.000$), and rewatched videos before exams (ave $= 3.87$, $p = 0.002$). However, there was no general use of the textbook during the online lectures (ave $= 2.91$, $p = 0.789$).

In contrast to the flipped structure of the class, the students were much more positive about the active learning components in the classroom. There was broad agreement with the following statements:

- Solving problems in class was helpful for my understanding of the topics (ave $= 4.35$, $p = 0.000$);
- I enjoyed discussing conceptual questions in class that were asked in the lecture videos (ave $= 4.17$, $p = 0.000$);
- The additional discussions and clarifications in the classroom sessions were important to understanding the material (ave $= 4.35$, $p = 0.000$);
- Working in groups during class helped me understand the material better than working on my own (ave $= 3.65$, $p = 0.025$).

The "working in groups" statement got the lowest rating of these questions relating the active classroom component of the course, but even those students who disliked the flipped class model (responding with a "1" or "2" to I enjoyed this flipped class-room) still were statistically neutral on this statement. Thus even students who did not like the flipped model did not feel that working in groups in class was worse than working alone. Moreover, in response to all the other statements, even the students who disliked the flipped class model agreed more than disagreed that the other components of the active classroom were beneficial to their learning.

5 Graduate Student Performance: Evidence from the Qualifying Exam

The survey shows that the active learning components of the courses were viewed by students as beneficial for their understanding of the material, but is there any real evidence that this was the case? Evaluating the actual impact on student learning of these active engagement techniques is very difficult without controlled studies, which is difficult to do when a course is taught once per year and there are 10–20 students in a class. However, we do have one piece of evidence regarding the impact on graduate students of active learning in the Classical Mechanics course, namely performance on the departmental qualifying exam.

The UT Arlington Physics qualifying exam is taken over two days at the beginning of the fall and spring semesters. It has four separate sections: Quantum Mechanics, Statistical Mechanics, Classical Mechanics, and Electromagnetism. The problems for these sections are each developed by a committee and cover upper division and beginning graduate student material. There are three possible outcomes. A student can pass a section (>60% score), fail (<40%), or score well enough to be eligible for an oral examination on the topic (>40% but <60%). A student who fails any of the four sections twice will be dismissed from the Ph.D. program. First-year students take the exam at the start of the spring semester after they have had Quantum Mechanics, Classical Mechanics, and Electromagnetism in the fall (they take Statistical Mechanics in the spring), and any exams that they did not pass are repeated at the start of the fall semester (the second year for the students). If a student does not pass a section after two tries, but scores well enough to merit an oral exam, the student's research advisor may request an oral exam on the topic for the student, which must be passed to remain in the Ph.D. program.

Since the Classical Mechanics course has used an active learning environment, none of the 37 physics graduate students having taken the course has failed the Classical Mechanics section of the qualifying exam. Two students out of 37 did not score an outright pass on the first attempt (they scored in the oral exam range). One of those two passed the section on the second attempt, while the other voluntarily left UTA before taking the section a second time. No other section of the qualifying exam has a similar record. They all have students who fail the first attempt, sometimes the second, and each year there are students who require an oral exam in some topic. Before the change in the course structure the pass rate for the Classical Mechanics portion of the qualifying exam was the same as for the others. Moreover, the problems that the Classical Mechanics committee uses have not varied much over the past decade since the exam tends to be constructed from a set of existing problems, some of which are directly taken from Chaps. 1 and 2 of the graduate textbook. While this evidence is perhaps more anecdotal than not, it is still supportive of the view that graduate student understanding of Classical Mechanics is somewhat more robust than other areas as a result of the introduction of active learning into the graduate course.

6 Summary and Conclusions

Beginning physics students are clearly novices (National Research Council 2000), but at some point in their education they begin to develop real expertise as they become physicists. By the time they are in their upper division courses, and certainly when they are graduate students, they begin think of themselves as belonging to the physics enterprise, as being physicists. Faculty often view graduate students as having developed a robust expertise in physics, though they need to deepen that expertise through an additional round of graduate coursework. Yet it is true that many upper division undergraduates still carry misconceptions and unresolved contradictions in their understanding that can persist into graduate school. Therefore, the kind of active learning techniques that are often deployed in lower division courses have a place in upper division and graduate courses.

In this paper, we describe an experiment that was done with graduate students to gauge their response to an active learning classroom. The student response was quite positive, and this experiment later motivated a restructuring of one of the core graduate courses, Classical Mechanics. After one semester, the course was further modified to utilize the flipped classroom model. A year later the same model was applied to an undergraduate course that was already being taught using active learning approaches, including as concept questions embedded in lecture and collaborative problem-solving. The flipped classroom model is an innovation that has spread quickly over the past few years, but in postsecondary education it has been used primarily in introductory, lower division courses. In fact, this is true for most of the pedagogical innovations being used in college and university classrooms. The application of pedagogical innovations in upper division and graduate courses (and reports of such efforts) is much less common.

Student response to the flipped classroom model was mixed, with no strong average preference, but with strong individual preferences, for and against. Zappe et al. (2009), in a study of a flipped Junior-level Architectural Engineering class, found that students liked some aspects of the flipped class, but they did not want all of the lectures to be flipped. The students in that study indicated that 50% flipped lectures would be optimal. One thing that became clear in the application of the flipped classroom model to the undergraduate class is that a significant part of the effectiveness of the concept questions that had developed for use during lecture were lost since the students were not be able to dialog with each other about the question. Thus, the video lectures in the flipped class essentially took a step backwards toward a "sit and listen" experience and lost some of the active aspects that occur in class.

Moreover, more advanced students have different motivations than introductory students who just want to pass a physics class and move on to their real objective. Advanced students are beginning to see themselves are part of a discipline, or members of a community. They may desire a bit more traditional instruction in the classroom, while still appreciating an active learning approach to classroom instruction. Clearly in our data there were a number of students who were not fond of the fully flipped approach. The popularity of concept questions, and the loss of the interactivity in lecture that moving to a flipped model produces, is another issue. Perhaps a mixed model in which there is still a flipped component, but some lectures (with concept questions) are still done in class along with more extensive problem-solving would ameliorate this negative effect of the flipped model. This would also provide a better experience for students who do not like the flipped model, at least not all the time (following the findings of Zappe et al. (2009), who were also dealing with an upper division course).

The flipped model implies that the class time freed up by moving the lectures to an online component will be used for active learning. However, depending on how the time is used this may or may not be the case, so the active learning component can be seen as something distinct from the class structure. Our findings are that graduate and upper division undergraduate students by and large responded very positively to an active learning environment based on collaborative problem-solving and peer-instruction, even if they did not care so much for the flipped class model itself. The students reported that the discussion of concept questions and collaborative problem-solving was enjoyable and helpful to their understanding of the material. Some evidence in favor of this finding comes from the performance of students on the graduate qualifying exam, which suggests that their mastery of Classical Mechanics is a bit better than the other core subjects, and that this is related to the introduction of active learning into the Classical Mechanics course. These findings overall provide evidence that courses for graduate and upper division undergraduate students can benefit from the pedagogical innovations that are improving student outcomes in lower division courses.

Acknowledgements We acknowledge support from NSF grant DUE-0856796 and Department of Education grants P200A130247 and P200A090284.

References

Ambrose BS (2004) Investigating student understanding in intermediate mechanics: identifying the need for a tutorial approach to instruction. Am J Phys 72:453–459

Bishop JL, Verleger MA (2013) The flipped classroom: a survey of the research. ASEE Conf Proc 30:1–18

Crouch CH, Mazur E (2001) Peer instruction: ten years of experience and results. Am J Phys 69:970–977

Deslauriers L, Schelew E, Wieman C (2011) Improved learning in a large-enrollment physics class. Science AAAS 332:862–864

Finkelstein ND, Pollock SJ (2005) Replicating and understanding successful innovations: implementing tutorials in introductory physics. Phys Rev ST Phys Educ Res 1:010101

Halloun IA, Hestenes D (1985) The initial knowledge state of college physics students. Am J Phys 53:1043–1055

Lopez RE (2008) Space physics and the teaching of undergraduate electromagnetism. ASR 42:1859–1863

Lopez RE, Gross NA (2008) Active learning for advanced students: the center for integrated space weather modeling graduate summer school. Adv Space Res 42:1864–1868

Mazur E (1997) Peer instruction. Prentice Hall, Upper Saddle River, NJ, pp 9–18

McDermott LC (1993) Guest comment: how we teach and how students learn—a mismatch? Am J Phys 61:295

McDermott LC (2001) Oersted medal lecture 2001: "Physics education research—the key to student learning". Am J Phys 69:1127–1137

McDermott LC, Shaffer PS (1992) Research as a guide for curriculum development: an example from introductory electricity. Part I: investigation of student understanding. Am J Phys 60:994–1003

National Research Council (2000) How people learn: Brain, mind, experience, and school: Expanded edition. National Academies Press

Otero V, Pollock S, Finkelstein N (2010) A physics department's role in preparing physics teachers: The colorado learning assistant model. Am J Phys 78:1218–1224

Perkins K, Adams W, Dubson M, Finkelstein N, Reid S, Wieman C, LeMaster R (2006) PhET: interactive simulations for teaching and learning physics. Phys Teach 44:18–23

Singh C, Belloni M, Christian W (2006) Improving students' understanding of quantum mechanics. Phys Today 59:43

Tucker B (2012) The flipped classroom. Educ Next 12:82–83

Tune JD, Sturek M, Basile DP (2013) Flipped classroom model improves graduate student performance in cardiovascular, respiratory, and renal physiology. Adv Physiol Educ 37:316–320

West DC, Pomeroy JR, Park JK, Gerstenberger EA, Sandoval J (2000) Critical thinking in graduate medical education: a role for concept mapping assessment? JAMA 284:1105–1110

Zappe S, Leicht R, Messner J, Litzinger T, Lee HW (2009) "Flipping" the classroom to explore active learning in a large undergraduate course. American Society for Engineering Education

Development of Preservice Teachers' Sense of Agency

Michael M. Hull and Haruko Uematsu

Abstract High school physics teachers face myriad restrictions and challenges that may make it difficult to teach using research-based methods. Whereas greater learning gains are found with active learning, this generally requires more time than rote lecturing. National standards which call for wide content coverage make such approaches prohibitive. This and other pressures may lead new teachers to feel that they have little control over what and how they teach. We think that it is important to build in teachers a sense of agency for their teaching, and we report on a curriculum for preservice teachers that aims to accomplish this.

1 Introduction

Many people think that the ability to teach is an artistic trait that you are either born with or not. However, physics education research has found that students tend to learn more in interactive engagement classrooms than in traditional lectures, regardless of how talented a lecturer their instructor may be (e.g., Hake 1998). This finding demonstrates that there *is* a science behind teaching. However, the *degree* to which teaching should be regarded as an art and the *degree* to which it should be treated as a science is one in which teachers and researchers have not reached a consensus. Sawyer discusses some curricula, such as *Success for All*, which "provide word-for-word scripts that teachers are strongly encouraged to follow" in order to "teacher-proof" instruction (Sawyer 2004). Considering as well the exhausting course load that high school teachers in many countries face, there are sentiments that it would

M. M. Hull (✉)
University of Vienna, Austrian Education Competence Centre Physics, Porzellangasse 4, 1090 Vienna, Austria
e-mail: michael.malvern.hull@univie.ac.at

H. Uematsu
Department of Physics, Tokyo Gakugei University, Nukuikita-Machi 4-1-1, Koganei-Shi, 184-0015 Tokyo, Japan

© The Editor(s) (if applicable) and The Author(s), under exclusive license to Springer Nature Switzerland AG 2020
J. Guisasola and K. Zuza (eds.), *Research and Innovation in Physics Education: Two Sides of the Same Coin*, Challenges in Physics Education, https://doi.org/10.1007/978-3-030-51182-1_17

209

be best to have teaching physics purely as a science, with minimal innovation by the teacher. With this model, the teacher would faithfully utilize curriculum developed by education researchers without a need for taking the time to interpret or understand the rationale behind the curriculum. Many education researchers, however, argue that responsive teaching is crucial (e.g., Sawyer (2004), Debarger et al. (2017), Harlow 2010, 2009, Robertson (2018), Schrittesser (2013)). Sawyer, for example, writes "...the best teachers apply immense creativity and profound content knowledge to their jobs, both in advance preparation and from moment to moment while in the classroom" (Sawyer 2004).

One reason why creativity is desirable in a teacher is that sometimes students respond to a prompt differently than expected by the creators of the reformed curriculum, and a creative teacher can react appropriately. An example of this is described in detail by Harlow (2010) who looks at two teachers utilizing a lesson from Physics and Everyday Thinking (PET). In studying magnetism, the worksheet has students draw a sketch of what they think is different inside of a nail that has been magnetized (by rubbing it with a magnet) from one that is not magnetized. The worksheet then has students predict what their picture would imply should happen if they cut the nail in half and each end of each half is held up to a magnet. After conducting the experiment, students are told to revise their models if necessary.

The curriculum developers expect students to think (incorrectly) that north poles accumulate at one end of the magnetized nail and south poles at the other end, as shown in Fig. 1. When cut in half, the developers predict students to (incorrectly) say that each end of the negative half of the nail will be attracted to the north end of the magnet. When students find that, in fact, one end of the half is attracted but the other end is repelled, students can revise their model to the correct one, that there are magnetic dipoles throughout the nail that are all aligned when the nail is magnetized. In both the classroom of Ms. Shay and the classroom of Ms. Carter, there was a group of students who did not have this specific incorrect explanation of what a magnetized nail looks like. Rather, they thought that the magnetic property comes about because magnetic dust had accumulated on the nail when it was rubbed by the magnet. Ms. Shay, staying true to the worksheet, had the students proceed to cut the nail regardless. The lesson did not succeed, because the results did not challenge the students' model. Ms. Carter, on the other hand, deviated from the worksheet productively. Namely, she had the students "wipe off" the magnetic dust. Seeing that the nail was still magnetic, the students revised their model.

Fig. 1 Expected student models for unmagnetized (left) and (incorrect) magnetized (right) nails

Unmagnetized nail

Magnetized nail

2   Curricular Knowledge and Perceived Agency

To respond as Ms. Carter did in this anecdote, we posit that a teacher must both have both knowledge about the intentions of the curriculum and a sense of agency. We elaborate upon each of these in turn.

2.1   Curricular Knowledge

Harlow describes the intervention of Ms. Carter as likely being heavily influenced by a professional development course she took that involved the same PET magnetism activity that she then had her own students carry out. In this course, she learned why the worksheet has students cut the nail in half, and that the overall point of the lesson is to help students in model building (Harlow 2009). Citing this example, Robertson argues that there is a need "to develop curricular knowledge—in particular, knowledge of the purposes of particular questions or sequences of questions within the curricula [teachers] use" (Robertson 2018). Such knowledge is important not only for knowing when to deviate from curriculum, but for other aspects of effective physics teaching as well, such as finding connections between physics topics so as to teach the subject as a coherent whole (Schneider and Krajcik 2002). Teachers do not, by default, perceive automatically the curricular knowledge embedded in the curriculum they implement (Robertson 2018; Ball and Cohen 1996; Biesta et al. 2015). Rather, curricular knowledge is something that must be learned (Robertson 2018; Schneider and Krajcik 2002; Ball and Cohen 1996; Davis and Krajcik 2005).

Just as education research has shown that physics students learn physics better when interactive engagement methods are used, it seems the same is true for teachers learning about curricular knowledge (Schneider and Krajcik 2002; Turpen et al. 2016). Schneider and Krajick conducted a study where several physics teachers were provided with curricular knowledge in the form of "overviews" to accompany the instructor's guides. Only one of the teachers mentioned actually reading the overviews (Schneider and Krajcik 2002). In the case of Ms. Carter, it was necessary for her to not only read the curricular knowledge, but also to understand it well enough to be able to act upon it in the moment with her own students. Turpen et al. write "a dissemination approach alone is insufficient for helping faculty to learn how to flexibly adapt instructional strategies based on students' reasoning and engagement" (Turpen et al. 2016). In arguing instead for an approach of "developing reflective teachers," they found it important to have the teachers in a professional development workshop to discuss with each other and undergo pedagogical sensemaking. Organized active learning about curricular knowledge that leads to positive results can take various forms (in addition to that described by Harlow), including a class where student teachers "dissect" curricular materials and discuss what curricular knowledge they think are contained (Robertson 2018) or in-service workshops where teachers are involved in co-creating new curriculum with education researchers (Debarger et al. 2017; Severance et al. 2016).

Even if a teacher understands curricular knowledge sufficiently to be able to apply that knowledge in the classroom setting, what ultimately matters is whether or not the teacher actually does apply it. Biesta et al. write that many teachers are faced in the school with a "mishmash of competing and vague ideas—personalization, choice, learning, subjects, etc.," "are regularly left confused about their role," and hence tend to think more about short-term obligations and less about the long-term purposes of education (Biesta et al. 2015). Considering this in light of the example from PET, we can imagine a teacher who thinks something like "cutting the nail won't challenge this group's model, and that is the point of the exercise, so there is kind of no point in having them do that. But if I go in some other direction and improvise, it might take up more time, and we have a tight schedule. It's the job of the curriculum developers to make the materials, and my job to follow them and then move on to the next thing." In addition to curricular knowledge, teachers need also a sense of agency to improvise as Ms. Carter did.

2.2 Perceived Agency

Although the concept of "agency" has been extensively discussed, particularly in sociological literature, relatively little has been written about teacher agency (Biesta et al. 2015). By "agency," we mean something very similar to what Enghag and Niedderer describe as "ownership," which "refers to the importance and need… to actually participate by discussions, choice, responsibility, and decision taking" (Enghag and Niedderer 2008). Milner-Bolotin (Milner-Bolotin 2001) describes "ownership of the learning process" for learners in project-based-instruction group projects in terms of three overlapping realms, with full overlap corresponding to most ownership: (1) does the learner find personal value in the project (e.g., finds the knowledge to be useful and/or connected to prior knowledge?); (2) does the learner feel in control to make decisions and be proactive?; and (3) does the learner feel responsible for the learning process and results of the project? We mean something very similar by "perceived agency" of a teacher, but we avoid the word "ownership" as, when used in the context of teaching rather than learning, it conjures up images of strict disciplinarian and/or traditional lecture-based forms of teaching.

A sense of agency not only provides teachers the "permission" needed to make productive in-the-moment changes to the curriculum, but also to use interactive engagement curriculum in general, especially when it might be inconvenient to do so. High school physics teachers graduating from Tokyo Gakugei University have reported feeling pressure to teach in the traditional style used by other teachers at their school instead of with the research-based curriculum they learned as preservice teachers. In many countries, the class time necessary to cover the wide breadth of topics put forth by national standards makes the use of research-based curriculum prohibitive, as interactive engagement typically requires more time. A teacher without a sense of agency may feel that he or she does not have the freedom to utilize such curriculum. Teachers are influenced not only by what administrators say, but also by parents and their perceptions of their students (Ball and Cohen 1996). For example, teachers may believe that students prefer traditional teaching, and this might prevent them from using interactive engagement methods (Biesta et al. 2015). We conceptualize a teacher with a strong sense of agency as one who, amongst a backdrop of various external influences, nevertheless perceives control of what is taught and of how it is taught, and we define "perceived agency" in this way. That is, **we define "perceived (teacher) agency" to be "a feeling of being in control over what is taught and of how it is taught."** A teacher with a weak sense of agency, on the other hand, feels controlled in this regard, either by the education system, student expectations, or the curriculum itself.

Although we have described curricular knowledge as a separate construct from perceived agency, we hypothesize that there is a connection between them. Specifically, we hypothesize that teachers with a strong sense of agency might be more inclined to seek out the curricular knowledge and that, conversely, teachers who effectively learn curricular knowledge might come to feel more agency.

3 Methodology

This report discusses our first steps in the development of a survey to measure teacher's sense of agency as well as preliminary findings from the administration of this survey. There are certain inherent limitations to such a device that must be addressed. Biesta et al. write that "agency is understood as an emergent phenomenon of actor-situation transaction," and, accordingly, utilize an ethnographic approach, interviewing only a few teachers and administrators about their teaching beliefs (Biesta et al. 2015). We agree that greater depth of exploration is afforded by such qualitative methods, and we feel that a study of agency cannot be limited to survey results. Firstly, surveys do not describe what the respondent actually does in practice. For this reason, our survey results can only describe the respondents' perceived agency. Even with this caveat, surveys inherently do not account for the context-sensitivity of the respondents' views. That is, it is easy to imagine a teacher answering a question like "Do you feel like you have control over the learning of your students?" differently just before and just after a class, particularly if a carefully planned lesson turns out to be a disaster. The benefit of a survey, however, is that, in principle, data from a large number of respondents can be obtained with relative ease.

Although a search online revealed no surveys to measure teacher agency specifically, two surveys were found which measure agency in other contexts, and these served as a basis for the survey we created. The first survey, the Ownership Measurement Questionnaire, was created by Milner-Bolotin to measure the feelings and beliefs of non-science majors who worked on a group project for a physical science course (Milner-Bolotin 2001). The second survey, Perceived Choice and Awareness of Self Scale, includes questions which measure whether or not one perceives a sense of choice behind his or her actions (Perceived Choice and Awareness of Self Scale (PCASS)). Both surveys utilize a five-point Likert scale, as does the survey we created. Since some items on these surveys were similar, they were condensed into single (instead of near duplicate) items. Questions were added based upon anecdotal conversations of the second author with graduates from the teacher training program at Tokyo Gakugei University (TGU). We wished to have an equal number of positive as negative statements, so this led to the creation of additional statements. Finally, to not have the survey be too burdensome, items that we predicted to not be useful and/or were redundant were removed. For example, we first modified the prompt "I have a sense of ownership of the group project I am working on" from the Milner-Bolotin survey to be "I have a sense of ownership of my teaching" and, conversely "I do not have a sense of ownership of my teaching." We then considered that students would likely either be confused by these prompts or that it would be redundant with the responses to the other prompts, and so we removed both prompts.

The final 38-item survey administered to preservice teachers (PSTs) at the start of the 2018 summer semester at the University of Vienna (UV) and during the 2018 spring semester at TGU consisted of the prompts in Table 1. The prime beside item numbers designates that the item is a paired statement for the un-primed statement preceding it. A "disagree" ("agree") on an un-primed (primed) statement was interpreted to denote a sense of agency. For example, if a PST disagreed with statement 5, "Curriculum is developed by experts, and it should be used without messing it up" (indicated by circling either "1" or "2"), then we considered that to be a demonstration of perceiving agency. If he or she agreed with statement 5', "Curricular resources are a guide for teachers to use or modify creatively, as the situation requires" (that is, they circled either "4" or "5"), then that was also taken to demonstrate perceived agency. The "b" next to the numbers of some prompts denotes that they are intended to capture the same sense as the corresponding questions without the "b". They are included to eventually check for consistency with the corresponding questions as one measure to validate the survey.

In the administration of the survey, the order of the questions was changed to separate the paired questions and the numbering was removed. At the start of the semester, students were instructed to complete the survey individually by drawing a circle around their response (the administered surveys are available upon request).

After students had completed the survey at UV (20 min were allowed), four of the prompts (Q.1, Q.2, Q.5, and Q.6) were discussed first in pairs and then in whole-class discussion. These prompts were chosen because they are well-known deterrents for using reformed curriculum.

This discussion was video-recorded, prior to which PSTs signed consent forms. Snippets of conversation are presented below to corroborate with the survey results, and some student comments were used to improve the survey. For example, some students interpreted the word "curriculum" in Q.5 to mean "national syllabus." Since the intent of the question was instead to ask about material which might be used to satisfy the national syllabus (such as PET or a textbook), this question was replaced with five new questions (the first five questions in Table 2). Furthermore, based upon this class discussion, Q.2, Q.2', Q.2b, and Q.2b' were replaced with Q.2*, Q.2'*,

Table 1 Questions of the agency survey administered pre-instruction at UV, summer semester 2018

1	If the principal of my school tells me to teach in a certain way, I will do my best to teach that way, even if I don't really want to
1'	I will teach in the way I think is best, regardless of what my principal or other teachers might think
1b	I will likely use whatever curriculum the teacher before me used at the schools where I will teach. I don't want to cause any trouble
1b'	It might be the case that at my school where I am teaching, a more experienced teacher will not want me to use research-based pedagogy but to instead stick to traditional ways of teaching. Nevertheless, I will keep trying to introduce curriculum that I think will be the most effective
2	I am required to teach a wide range of content to my students. I do not have time to help students understand a given topic as deeply as I might like
2'	If my students do not understand what they are learning, I will take more time with the material, even if that means that some planned topics are not taught during the year
2b	Sometimes teachers have to teach in a way that is not very effective because national standards require it
2b'	In the case where national standards ask us to teach in a way that is ineffective, I will ignore the standards and teach in my own way
3	I prefer curriculum that tell the teacher exactly what to do, so that I don't risk making the wrong decision
3'	I would welcome national curricular reform that puts more responsibility on the teacher to make decisions in the moment in response to what students say or do
4	Education researchers know what's best for student learning. I will try to teach exactly as they suggest
4'	Education researchers might know how most students learn, but not specifically my students. I will therefore think carefully before using a new research-based curriculum about whether it is appropriate for my own students or not
5	Curriculum is developed by experts, and it should be used without messing it up
5'	Curricular resources are a guide for teachers to use or modify creatively, as the situation requires
6	My students will have taken many classes before taking my class, and they will have an idea of how a class "should go". I need to teach in that style too, otherwise it will be too strange for my students
6'	It is OK if my teaching style is different than what my students are used to. They will figure it out
7	I need to listen carefully to the demands of the parents of my students to make sure I'm teaching what they want their children to learn
7'	Parents should not tell me what or how to teach—I am the expert, not them
8	Teaching is just a job so I can get a paycheck—there is no benefit to me beyond that
8'	I find personal value in teaching
9	The content I teach and the way that I teach it are not something for me to decide

(continued)

Table 1 (continued)

9'	I feel that I have control over what I teach and how I teach it
9b	Generally, someone else decides what and how I teach
9b'	Generally, I choose what and how I teach.
10	It doesn't really matter whether I do my part in helping students learn or not—they will meet plenty of other teachers
10'	I feel responsible for doing my part in helping my students learn.
10b	Students will learn if they want to and won't if they don't want to—their learning is not my responsibility
10b'	I feel a personal responsibility for the learning of my students
11	If my colleagues have gone home for the day, I will go home too, even if the quality of my lessons suffers
11'	I will provide quality education to my students, even if I need to spend more time preparing for class than my colleagues do
12	What my students learn in my class will have little benefit for them in other courses and/or in everyday life
12'	I think what my students learn in my class will be useful for them in other courses and/or in everyday life
12b	The skills my students learn in my class, if any, will have little benefit to them once they graduate from school
12b'	I think the skills my students learn in my class will help them to succeed in the future
13	I think the progress of my students is independent of anything I as a teacher might do
13'	I think I have control over the progress of my students
14	It is not my job to make students think deeply—that is their decision
14'	I feel responsible for making my students think deeply

A Japanese version was administered at TGU during the spring semester.

Q.2b*, and Q.2b'* (see Table 2). These 9 new questions were administered during the second week of class as a follow-up survey at UV. By this point, four students had already been assigned work that could influence their answers to these 9 questions. As such, their responses to these 9 questions were not considered in analyzing student growth pre-/post-semester. The posttest survey administered at the end of the summer semester at UV included these 9 new questions as well as the questions that were not modified from the pretest.

We now present preliminary findings from this survey in the context of addressing two questions:

(1) Is there a difference in the perceived agency of preservice teachers (PSTs) at Tokyo Gakugei University (TGU) and PSTs at the University of Vienna (UV)?
(2) Does perceived agency of PSTs change with increased curricular knowledge?

Table 2 Questions created based upon discussion with students regarding their responses to the agency survey

15	I will probably just use whatever physics textbook the teacher before me used. If it worked for her, it will work for me
15'	I will consider carefully what physics textbook to use in my classroom
16	Once I choose a physics textbook for my classroom, I will follow it carefully
16'	Once I choose a physics textbook, I will just use it as a guide. I will not hesitate to skip sections or point out to students which parts I think are poorly-worded, confusing, or wrong
16"	In my physics class, I will combine textbooks and other materials, taking the best from each source
2*	I am required to teach a wide range of physics content to my students. I do not have time to help students understand a given topic, like Newton's laws, as deeply as I might like
2'*	If my physics students do not understand what they are learning, I will take more time with the material, even if that means that some planned topics are not taught in class
2b*	Sometimes physics teachers have to teach in a way that is not very effective as a result of national standards
2b'*	In the case where following national standards would mean teaching physics in a way that is ineffective, I will ignore the standards and teach in my own way

These questions were administered at UV as a follow-up survey.

4 Study Design

As a preliminary pilot study for the survey described above, it first was administered to only a small sample of PSTs both at TGU (in Japanese) and at UV (in English). At TGU, the sample consisted of 27 undergraduate PSTs in an introductory seminar course majoring in physics education, 26 of whom were 1st-semester students, as well as 13 teaching assistants majoring in physics education, 11 of whom had BS degrees, and 2 of whom were in their 4th year for the BS degree. The data sample from UV consists of the 16 PSTs, all of whom majored in physics education, enrolled in a seminar course taught by the first author. This course utilized Open Source Tutorials (OSTs), guided worksheets developed at the University of Maryland (Elby et al. 2007). Similar to PET, OSTs imbed within them research-based assumptions about student patterns in reasoning, and so curricular knowledge is important to recognize when students do not respond as predicted. Weekly homework assignments had PSTs at UV predict the intentions of the tutorial they had gone through earlier in class that week (i.e., to try to discern the curricular knowledge in each tutorial), and they checked their predictions in the following class when they received the instructor's guide. This style of the assignment was given to students with the intention of helping them to develop curricular knowledge.

5 Analysis and Results

For each question on the agency survey, the number of codes expressing perceived agency was divided by the total number of responses for that item. In cases where PSTs left an item blank (as they were instructed to do if they did not understand the item), the denominator is hence decreased. In Figs. 2 and 3, the x-axis contains the question number and the y-axis is the percentage of "agency" codes out of total number of responses for that question. As seen in Fig. 2, for most items, UV PSTs at the beginning of the semester have a higher perceived agency score than the TGU PSTs, but not for all items. Averaged across all items, UV PSTs have a score of 77%, in comparison with the 53% for TGU undergraduate students or the 57% of TGU graduate students.

In some regards, these survey findings are not surprising. The observation made by the second author that TGU graduates face pressure to follow the status quo and not to use research-based curriculum is visible in this data as well. Namely, for Q.1', roughly 25% of TGU PSTs agreed with the statement "I will teach in the way I think is best, regardless of what my principal or other teachers might think." A noticeably large difference between the UV PSTs and the TGU PSTs are visible for Q.7 and Q.7', regarding taking into account the views of student parents when teaching a lesson. Considering the growing concern of teachers in Japan about "monster parents," this is also not surprising.

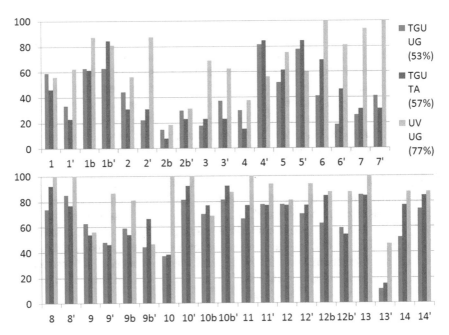

Fig. 2 Comparison of new PSTs at TGU (blue, left), older PSTs at TGU (red, middle), and PSTs at the start of the seminar course at UV (green, right)

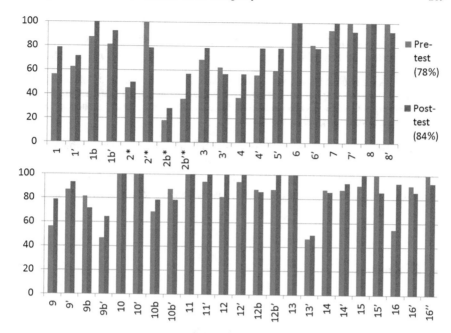

Fig. 3 "Pretest" questions are identical to those of the posttest, taken from the surveys given on the first day of class and the following week

Overall, it seems that UV PSTs report feeling a rather strong sense of agency, a finding that is also noticeable in recorded data from class. After PSTs had completed the survey, the first author had them discuss certain questions with each other and then to share their views with the whole class. In response to Q.1, "If the principal of my school tells me to teach in a certain way, I will do my best to teach that way, even if I don't really want to," Adler (all PST names are pseudonyms) replied

> I put "disagree", because I feel, if I don't think that this is the best way to teach my students, how should the students get the value of it? If I myself disagree with the subject I'm teaching? For me it's hard to teach something if I'm not believing in it, not really thinking "this is making sense, this is a good way of teaching it", for me it is hard to provide that. I think it is like lying to yourself. But I only put "disagree" there and not "strongly disagree", because, at the end of the day, the principal is still our boss; we are not working independently. So I am under them, and, of course, they are the boss, they need to inform what I do in my classroom.

When asked a modified version of Q.5, "Research-based textbooks are developed by experts, and they should be used without messing it up," two students responded

> Billie: Most textbooks I studied in the last few years had extreme faults in them, so it's not that they know the answer to everything, even if they are good. It's not always possible to find a perfect textbook, so you have to mess with it a little bit, because every class is different, so you can't do exactly what they would have you do.

> Dresden: I think it is seldom a good idea to use just one source for those kinds of things, so probably you should take multiple books and maybe compare or whatever.

Finally, in response to Q.6, "My students will have taken many classes before taking my class, and they will have an idea of how a class 'should go'. I need to teach in that style too, otherwise it will be too strange for my students," Billie responded

> I mean, it's really difficult, just saying that you have to do it the same as the other teacher because I'm not another teacher, I'm me, and I have to stay me even if I have a class that is not used to how I teach. If I tried to be someone else, the class wouldn't be authentic. So I don't think it would be good for the class.

Taken collectively, the utterances of these PSTs indicate that they are thinking that they will rely upon their own views about what is the best way to teach and that they will be proactive in deciding which curriculum to use and how to implement it. As shown in Fig. 3, the results of our survey are consistent with a growth in perceived agency.

6 Discussion

We are interested in addressing two main questions in our work:

(1) Is there a difference in the perceived agency of preservice teachers (PSTs) at Tokyo Gakugei University (TGU) and PSTs at the University of Vienna (UV)?
(2) Does perceived agency of PSTs change with increased curricular knowledge?

At this point, our data is too scarce to be able to reach conclusions on either of these questions. Each bar in Figs. 2 and 3, for example, contains fewer than 20 data points. Furthermore, although we presented unweighted averages across all survey items above, this was done only to guide the eyes in noticing that some bars across items are higher than others. We certainly do not mean to imply at this point that each item indicates equally strongly that the respondent has a sense of agency. Finally, we do not assume that all survey questions are functioning well. For example, the striking difference in responses to Q.10 in comparison with Q.10' in Fig. 2 suggests that the statements might not be interpreted as intended. We do think, however, that these pilot study results warrant additional validation of the survey items and administration to additional PSTs in subsequent semesters.

We have begun conducting survey validation interviews to improve the survey. Once we have completed this, we plan to continue to administer the improved survey at both TGU and UV. Once enough data is accumulated, we plan to evaluate survey validity by other means as well, such as by calculating Cronbach's alpha to measure internal consistency of the survey items. Once this is done, we plan to administer a final version of the survey to accumulate sufficient data to see if differences in data sets are statistically significant or not. Until then, all we can do is use the data available to speculate on what it might be demonstrating.

6.1   Relation Between Curricular Knowledge and Perceived Agency

Although we defer to future research to measure curricular knowledge of PSTs and to check for correlation with perceived agency as measured by the survey, there are two pieces of data from our present study to suggest that there might be a connection between curricular knowledge and perceived agency.

First, from Fig. 3, we see that the item with the most dramatic change was Q.16: "Once I choose a physics textbook for my classroom, I will follow it carefully," which saw an increase of 40% of "disagree" and "strongly disagree" statements. This is one of the questions specifically created to relate to curricular knowledge. We find it plausible that the increased score with this item is due at least in part to PSTs becoming more predisposed to search for curricular knowledge, a habit that was likely built by doing exactly that continuously throughout the course.

Second, at the end of the semester, UV PSTs were given the same instructions they had seen every week for their homework to dissect a tutorial, only this time with a tutorial they had never seen before. They were also asked these questions:

(1) *Do you think going through this dissection process is a worthwhile thing to do (your answer is not being graded, so please be honest!)? If so, what is the benefit of doing this process?*

(2) *What effect, if any, do you think doing such a process as this might have on your future teaching?*

Some responses clearly associated this process with a sense of agency:

Karsten: The benefit for me is to decide whether I would use a tutorial like this or if it needs to be "upgraded" in a way I want to use it in my classes.

Gerry: I do think that this process is useful because you have to think about what may go wrong (helpful especially for teaching in school) and what purpose certain exercises serve.

Adler: I think I'll definitely need to do such a process in my future as a teacher, because we decide what kind of exercises students will have to do, and therefore I'll always have to watch out for different materials and ask myself about the purpose of these and if they really impart values and knowledge the way I would like them to.

Taken collectively, these statements suggest that at least some of the PSTs are recognizing that searching for curricular knowledge enables them to exercise their agency by deciding what curriculum to use and how to use it.

6.2   Is There a Difference Between PSTs at TGU and at UV?

The preliminary data presented in this report suggests that PSTs at UV have a stronger sense of agency than the PSTs at TGU. If, after further survey validation and data accumulation, this difference persists and turns out to be statistically significant, it is interesting to consider what might account for the difference. It would be interesting, for example, to consider differences in how rigidly physics lessons are dictated by national standards. It is well known, for example, that high school physics teachers in Japan must prepare their students for college entrance exams that are largely standardized and that carry much more weight than, say, in Austria.

In general, although some of the survey items explicitly place the PSTs in the context of a situation where they must go against the established structure, we suspect that the differences between the PSTs in Tokyo and in Vienna are themselves manifest from structural differences, for example, in the education system in which they prepare themselves to work. In line with how sociological literature has described "agency" in general, teacher agency should not be perceived as a sort of opposite to structure, but rather enabled by it (Sewell 1992; Smith 2013; Olitsky 2006). Sewell (Sewell 1992), quoting Giddens, writes

> Human agency and structure, far from being opposed, in fact presuppose each other. Structures are enacted by what Giddens calls "knowledgeable" human agents (i.e., people who know what they are doing and how to do it), and agents act by putting into practice their necessarily structured knowledge. Hence, "structures must not be conceptualized as simply placing constraints on human agency, but as enabling".

With this consideration in mind, we must also wonder what, if any, disadvantages exist for teachers of PSTs with particularly strong senses of agency. Is it best for new teachers, for example, to adamantly insist on teaching using the latest curriculum that they were trained in as a PST, or is it better to wait a few years to get the "lay of the land" before gradually attempting to induce changes? Concretely, is getting PSTs to disagree with Q.1, "If the principal of my school tells me to teach in a certain way, I will do my best to teach that way, even if I don't really want to," a worthwhile learning objective for PST trainers? Although strongly disagreeing with this statement is not necessarily a positive thing, we do see it as strongly indicating a sense of agency, which we perceive overall as improving instruction. Perhaps the ideal would be for PSTs to disagree with this statement, for example, because they would aim to discuss their rationale for teaching in a research-based manner with their supervisor instead of immediately consenting to "teach something I'm not believing in" (Adler).

References

Ball DL, Cohen DK (1996) Reform by the book: what Is—or might be—the role of curriculum materials in teacher learning and instructional reform? Educ Res 25:6–14

Biesta G, Priestley M, Robinson S (2015) The role of beliefs in teacher agency. Teachers Teach 21:624–640

Davis EA, Krajcik JS (2005) Designing educative curriculum materials to promote teacher learning. Educ Res 34:3–14

Debarger AH, Penuel WR, Moorthy S, Beauvineau Y, Kennedy CA, Boscardin CK (2017) Investigating purposeful science curriculum adaptation as a strategy to improve teaching and learning. Sci Educ 101:66–98

Elby A et al (2007) Open source tutorials in physics sense-making. DVD, funded by NSF DUE-0341447

Enghag M, Niedderer H (2008) Two dimensions of student ownership of learning during small-group work in physics. Int J Sci Math Educ 6:629–653

Hake RR (1998) Interactive-engagement versus traditional methods: a six-thousand-student survey of mechanics test data for introductory physics courses. Am J Phys 66:64–74

Harlow DB (2009) Structures and improvisation for inquiry-based science instruction: a teacher's adaptation of a model of magnetism activity. Sci Educ 142

Harlow DB (2010) Uncovering the hidden decisions that shape curricula. AIP Conf Proc 21–4

Milner-Bolotin M (2001) *The effects of topic choice in project-based instruction on undergraduate physical science students' interest, ownership, and motivation* University of Texas at Austin

Olitsky S (2006) Structure, agency, and the development of students' identities as learners. Cult Sci Edu 1:745–766

Perceived Choice and Awareness of Self Scale (PCASS) selfdeterminationtheory.org/pcass/

Robertson AD (2018) Supporting the development of curricular knowledge among novice physics instructors. Am J Phys 86:305–315

Sawyer RK (2004) Creative teaching: collaborative discussion as disciplined improvisation. Educ Res 33:12–20

Schneider RM, Krajcik J (2002) Supporting science teacher learning: the role of educative curriculum materials. J Sci Teacher Educ 13:221–245

Schrittesser I (2013) From novice to professional: teachers for the 21st century and how they learn their job. Learning to be a teacher in a changing World Barcelona

Severance S, Penuel WR, Sumner T, Leary H (2016) Organizing for teacher agency in curricular co-design. J Learn Sci 25:531–564

Sewell WH (1992) A theory of structure: duality, agency, and transformation. Am J Sociol 98:1–29

Smith JA (2013) Structure, agency, complexity theory and interdisciplinary research in education studies. Policy Futures Educ 11:564–574

Turpen C, Olmstead AR, Jardine H (2016) A case of physics faculty engaging in pedagogical sense-making. In: Proceedings of the physics education research conference. Sacramento

Is Participation in Public Engagement an Integral Part of Shaping Physics Students' Identity?

Claudia Fracchiolla, Brean Prefontaine, Manuel Vasquez, and Kathleen Hinko

Abstract Out-of-school (or informal) STEM experiences have played an important role in an era where the line between facts and opinions is blurred. Spaces for dialogue and exploration are key to improve public perception of science, and educators, including those in out-of-school spaces, are vital agents in improving those perceptions. More importantly, increasing the number and diversity of students who choose STEM fields has become a key objective nationwide, to guarantee that STEM fields reflect the rich diversity of the communities it serves. This has prompted a steady increase in the number of informal STEM education programs, resources, and public campaigns, mostly targeted at youth from underrepresented minorities. However, the impact that participation in these programs has on those who facilitate them is largely understudied. In this study, we seek to understand university students' negotiation of physics identity after they participate as facilitators in an informal physics program.

C. Fracchiolla (✉)
School of Education, University College Dublin, Belfield, Dublin 4, Ireland
e-mail: claudia.fracchiolla@ucd.ie

B. Prefontaine · K. Hinko
Department of Physics and Astronomy, Michigan State University, East Lansing, MI 48824, USA
e-mail: prefont4@msu.edu

K. Hinko
e-mail: hinko@msu.edu

M. Vasquez
University of Colorado Boulder, Boulder, CO 80309, USA

K. Hinko
Lyman Briggs College, Michigan State University, East Lansing, MI 48824, USA

J. Guisasola and K. Zuza (eds.), *Research and Innovation in Physics Education: Two Sides of the Same Coin*, Challenges in Physics Education,
https://doi.org/10.1007/978-3-030-51182-1_18

225

1 Introduction

Disparity of representation in STEM fields has stirred the scientific community to develop strategies to recruit members of underrepresented minorities. Increasing the number and diversity of students who choose STEM subjects is a key educational objective of most EU Member States, as well as in the USA (Peña-López 2016; Kearney 2016). Despite significant resources being made available to meet this objective, through targeted campaigns and public engagement requirements within most grants, lack of diversity and equity in STEM fields is a persistent issue (McGee and Bentley 2017; McGee 2016; Miner et al. 2018). Recruitment efforts alone will not be sufficient unless we identify what deters students from pursuing and completing careers in STEM. Factors such as student attitude, self-efficacy, sense of belonging, motivation, and identity are understood to be important (Lewis et al. 2017; Tellhed et al. 2017) in addressing the number, and diversity, of students who persist in STEM subjects. More specifically, studies (Irving and Sayre 2015; Hazari et al. 2013) show that a person's self-association with physics is the strongest predictor of a person's future career path involving physics. Thus, without changing cultural practices within the fields, existing efforts will not suffice to improve current underrepresentation in STEM (Hyater-Adams et al. 2018a).

Informal physics programs, also known as outreach or public engagement, are often designed to offer spaces to develop young people's interest and understanding, while addressing issues of representation by providing the opportunity to build participants' science identity, belonging, and sense of community (National Research Council 2009; Greenhow and Robelia 2009; Pattison et al. 2018). The authors' previous work (Anderson and Nashon 2016) has shown that the motivations of university students' who lead after-school physics programs changed from extrinsic to intrinsic factors—a key factor associated with robust physics identity formation. Personally, several of the authors' interest in studying physics was sparked through informal activities. Facilitating informal physics programs was a coping mechanism for dealing with roadblocks encountered through the PhD for several authors, including the first author who, as a Latina woman in STEM, experienced first-hand issues of equity and inclusion. Research, and the authors' own experiences, indicates that participation in informal learning activities has a positive effect on underrepresented groups, particularly women because these are spaces where everyone's ideas, needs, experiences, and backgrounds are often acknowledged (Hyater-Adams et al. 2018a, b; National Research Council 2009). Furthermore, research indicates that many of those wanting to facilitate these programs are from underrepresented minorities themselves, because of a desire to give back to the community and become role models (Hyater-Adams et al. 2018a, b; Collins 2002).

There are a growing number of studies focusing on understanding science identity (Hazari et al. 2013; Hyater-Adams et al. 2018; Carlone and Johnson 2007); however, most of these studies are set in formal learning environments. While structures exist (http://www.informalscience.org/evaluation) to assess the impact of informal learning on those who participate, little work has been done on determining the impact of participation in informal STEM programs on those who facilitate them, often consisting of undergraduate and graduate students in STEM. The US National Academy of Sciences cites limitations on such studies due to the variety of programs and the difficulty in applying research in practice (National Research Council 2009). In this study, we are looking to eliminate those limitations by developing mechanisms to assess impacts on facilitators using a framework to determine discipline-based identity formation. More specifically, we seek to determine if Wegner's community of practice (CoP) framework (Wenger 2010) is a viable tool to address the question: Is participation as educators in informal physics learning programs integral to the development of physics students' identity? We use previous work that has taken the physics community itself to be a CoP (Rodriguez et al. 2018; Close et al. 2016) to explore the implications of voluntary participation in informal programs through the lens of the CoP framework. We hypothesize that informal physics programs can also function as CoPs; thus, students pursuing degrees in physics while engaging in informal physics programs are then members of both CoPs. In this paper, we report on the efficacy of using the CoP framework as a lens on student experience in informal physics programs. We operationalize the theoretical framework for informal physics programs by looking at student interviews about their experiences as educational facilitators. Based on the analysis of several subjects, we suggest that certain informal programs may function as CoPs and provide meaningful physics experiences for physics students. Furthermore, we propose future work to apply the CoP framework more extensively to investigate programs.

2 Framework

Identity is often referred as a definition of one-self. From a Vygotskian perspective, identity is defined as a sociocultural construct that includes and is affected by everything that surrounds us and we consider ours (Holland et al. 1998). Holland et al. determined that our understanding of self and identity is dynamic, constantly subjected to change, being reassessed and molded by the environment and culture, and socially performed through interactions of individuals with groups and collective spaces.

A person's discipline-based identity, such as a physics identity, is deeply related to one's perceived self-association with the discipline, i.e., physics (Hazari et al. 2013). Considering that identity is a social construct, then an individual's physics identity can be mediated by sociocultural context, interactions, and resources available (Hyater-Adams et al. 2018; Brandt 2008; Rahm 2007). The community of practice (CoP) framework (Wenger 2010) informs how identity is created through practice as a social enterprise. CoPs are groups of people that collectively engage in a learning process and work toward achieving learning goals. In the community of practice all the members of the community share a drive to improve the community and they work collectively to create new, better approaches to reach their goal and move the community forward. Not every group or community constitutes a CoP—there are three main characteristics used to differentiate them: A CoP is a group that has some shared goal or expertise (domain) that helps each other achieve the goals (community) and that has a set of norms, repertoire, and shared information to achieve the domain (practices) (Wenger 2010).

There is a three-way connection between identity, practice, and community. By acknowledging the individual, the community is giving the individual a sense of belonging, which is associated with a person's identity; through the practice the individual negotiates how she/he participates in the CoP; and together the community develops and engages in the practices. Therefore, an individual's membership in the community determines their identity and self-association with that community. In the CoP framework, Wegner identifies five categories that link community of practice to identity. These categories determine how individuals develop their identity through participation in the CoP. See Table 1 for details.

3   Data Collection

In this work, we test the use of Wegner's five categories of identity to identify what practices, experiences, and interactions have an impact on students' discipline-based identity. Here we describe our application of the CoP framework to understand how students' identity as (1) members of the physics community and as (2) members of an informal physics program community was impacted.

In order to investigate the impact of participation, we designed a qualitative study in which university students actively participating in informal physics programs were interviewed. Interviews can provide a deep insight into the different factors that influence participants' sense of self-association with the physics field and informal physics programs. Interviews also allow us to examine students' perception of the communities and their perceived level of participation within the communities. We designed a semi-structured interview protocol that asked participants to discuss their experiences in the program and their perceptions toward physics and informal physics programs. Examples of the questions include: Why did you decide to volunteer for the program? Why would you (or would you not) volunteer again? Do you identify as a physicist? How did you end up in physics? The semi-structured interviews allowed us to have the freedom to follow up on questions and be able to capture as much of the narrative as possible. Interviews were conducted by two researchers and lasted about an hour on average.

In this study, we focused on interviews from participants in a US after-school physics program, Partnership for Informal Science Education in the Community (PISEC). PISEC is a semester long program in which PhD physics students teach inquiry-based activities to primary school children. The activities are based on a constructivist model for educational after-school environments and designed as exploratory leading activities involving intergenerational work. In PISEC, we referred to the volunteers as university educators (UEs). For our study, we conducted 15 interviews with UEs who had participated in the Spring 2016 semester of PISEC.

In addition, we asked the interviewees to complete a demographic survey at the end of the interview and to take field notes throughout their semester in PISEC. To probe if the CoP framework is suitable to study how participating in PISEC impacts UEs physics identity, we chose two interviews from the data set. The two interviews were randomly selected. Both subjects had the same role as facilitators (there is not a hierarchical system in PISEC for the volunteers, they all invest the same amount of time per week, per semester). One of the subjects is a female student, which we will call Sorcha, on her third year of graduate school. Sorcha, had some struggles starting her PhD, started her PhD in a different institution and dropout within the first year. She restarted her PhD a year later in the school where she completed her undergraduate degree. At the time of the interview, she had completed one semester of PISEC but

Table 1 Operationalized descriptions and example quotes from interviews with the subjects

Code	Description	Example
Community Membership (CM)	The forms of competence developed and valued by participants in the community, such as ways in which community members interact; perspectives and interpretations they share; use of a shared repertoire and resources, how we look at the world, how we relate to others, and what we know how to do	*There were a few different tours, and one of them had their teacher with them, their science teacher, and he was really helpful in- it was really informative to me to see how he took what I said and explained it to them. I was trying to make it accessible, but he really knew how to do that, so that was cool*
Learning Trajectory (LT)	Events that have taken place in the past or things that have been learned that resulted in the participant becoming a member of the community. Incorporates past identities and possible futures into making meaning of the present; i.e., experiences that have led them to participate in different forms in different CoPs. Participation in a community impacts an individual's identity only to the extent that the practice of the community incorporates that person's past and fits into a valued future. The learning trajectory influences what elements of participation are perceived as important and what are marginal	*So yeah, I wanted to do biology at first. I kind of realized that- the more I thought about it, I realized that the questions you ask and answer in biology aren't the kind I'm necessarily that interested in*

(continued)

Table 1 (continued)

Code	Description	Example
Negotiated Experience (NE)	Refers to the process of making meaning of experiences through participation in the community. It is related to our interactions with other members of the community and how those interactions form or perceptions of how individuals do or do not see themselves as members of the CoP	*Kind of, sort of. Honestly a lot of other people see me more as a scientist than I see myself as a scientist*
Nexus of multimembership (NM)	Negotiating being members of two or more communities and how their role serves as a bridge between those communities. Each individual is composed of multiple identities and how they negotiate membership to these different communities	*I don't think I will do physical education research but I do think that I will always to reach out to the non-scientific community and communicate science, so and try to do outreach with younger people*
Relationship between the local and global (RbLG)	Related to the sense of belonging. Individuals are constantly negotiating their local ways of belonging and how that fits to a broader spectrum of practices, styles, and discourses. Being a member of a local CoP is connected with being a member of the more universal community; for example, being part of a physics community in an institution and belonging to the community of physics at large	*Actually that really- yeah, that's actually great. We're at a university and we're talking about interacting with elementary, middle schoolers*

had previous experience in other informal science programs. The other interview is a male student, which we will call Eoin, in his first year of graduate school. At the time of the interview, Eoin had done two semesters of PISEC. If the framework proof to be a suitable tool to answer our proposed research question, then we will apply the framework to the complete data set.

4 Framework Operationalization

The CoP framework was operationalized based on the context of informal physics programs and physics communities of practice. In order to test the framework to study the mechanism around student physics identity formation within informal physics programs, narrative inquiry was used to understand participant experiences in physics and PISEC by looking at the content and language used in the interviews. Table 1 contains the codebook used to analyze the interviews, based on Wegner's framework.

Three researchers independently coded the interviews with the identity categories. This process helped us refine and validate our operationalized codes. Through the iteration process, we discussed and clarified the codes in which there were disagreements to further refine the codes' definitions. Interrater reliability was conducted between two researchers and then with the third researcher independently. All codes from the interviews were discussed, and discrepancies in the coding were resolved.

After the initial iteration of the identity codes, we realized that there was the need to add a set of subcodes for each category, to help us determine in which way the particular experience had an impact on the individual's participation within the community. Therefore, we added three subcodes (in, out, neutral). The *in* subcode refers to positive experiences that indicate a movement inwards to the community, and the *out* subcode refers to experiences that may have deterred the individual from further engagement and participation. Finally, the *neutral* subcode represents cases in which it was not possible to determine, through the narrative, if the individual considered that to be a positive or negative experience that would impact her/his participation.

5 Analysis

Participating in PISEC has a large impact not only on the children but also for the UEs that facilitate the activities. We hypothesize that the interactions UEs have with the children and peers in the program help mold their definition of self within their discipline. As once described by Cooley (1902) the looking-glass self-idea explains how individuals develop the image of self-based on how others, interpreting other's actions and responses to themselves, an individual continuously reevaluates their definition of self and changes based on these. In previous research conducted by the authors (Anderson and Nashon 2016), we identified evidence that interactions with the children and other members of PISEC prompted changes in their motivation for participation. The value given to participation went from mostly external factors to intrinsic ones. This change or reevaluation speaks about how the identity of the UEs is being impacted by their participation in PISEC through the interactions and experiences they have. We have expanded the initial work in this paper to explore the use of the CoP framework as a tool to understand how different experiences affect UEs participation in the Physics and PISEC CoP, and therefore their discipline-based identity. Our goal is to determine if, by using CoP framework, we can identify from the UEs' narratives what activities and experiences affect the level of engagement of the UEs in the PISEC and physics community. In this paper, we are focusing on determining if the framework is refined enough to identify differences in experiences the UEs had in the PISEC and physics community and how those experiences affected their membership in the respective community. In future work, we will apply the framework to how experience in one community may affect their membership in the other community.

5.1 PISEC

In Fig. 1, we noticed that there are clear differences in Sorcha's and Eoin's interactions and experiences within the PISEC CoP. This is evident from the coding analysis, in particular Eoin had statistically significant more negotiated experiences (NE) codes than Sorcha, which indicates that the external interactions, with either peers or children, had a bigger impact for Eoin than for Sorcha. For example, when asked about experiences in PISEC Eoin talked about working with the same two children each week: *[W]hen I worked with Shane and Paddy, I felt like, you know- Like I would come and there would be like a reunion, like hey guys, and they'd be like hey! You know, it was cute. Like when they came by the lab they indicated that they wanted to- I think Paddy asked if he could work at my lab when he got older or something. It was just pretty fun. Yeah, I definitely felt like an older brotherly connection with them or*

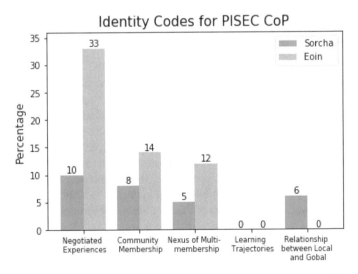

Fig. 1 Identity codes for the PISEC CoP for both subjects. Identity code values have been normalized with respect to the total number of identity codes (regardless of the community) within the interviews

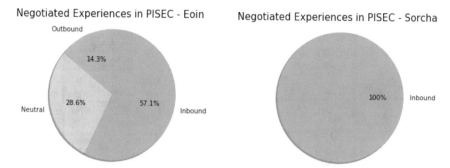

Fig. 2 Negotiated experiences subcodes for the PISEC CoP for both subjects. Values have been normalized with respect to the total number of NE codes within the interviews

something. This segment was coded as NE-inbound because it shows that he developed a deep connection with other members of the community, which reinforced his sense of belonging and therefore identity within the PISEC community.

However, if we look into the NE subcodes for the PISEC community, we noticed although Eoin has statistically significantly more NE codes than Sorcha, a fraction of those experiences (more than 10%) seemed to negatively contribute to his membership in PISEC (see Fig. 2). In contrast, all of Sorcha's NE experiences positively contributed to her PISEC membership.

5.2 Communities

The first thing we noticed is that there is a difference in the presence of codes for each of the communities between Eoin and Sorcha. For Sorcha, only 16% of codes are connected to the PISEC, while 31% are connected to physics; for Eoin, 34% are for PISEC and 21% for Physics. This difference can be a result of the fact that Sorcha is in her third year of physics graduate school and only her first semester in PISEC; therefore, she has more connections and stories related to the physics community of practice than PISEC, while Eoin is in his first year of PhD and second semester in PISEC, therefore having more connections in PISEC than Sorcha and maybe less in Physics. This difference could also be affected by the nature of the semi-structured interview protocol, because even though there are clearly defined questions, the interviewer is free to ask follow-up questions when particular responses seemed relevant to the research questions (Fig. 3).

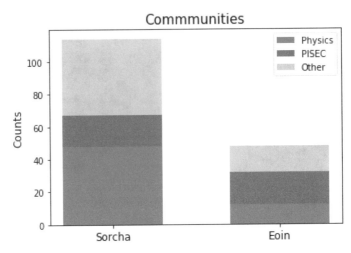

Fig. 3 Communities codes for both subjects. Values have been normalized with respect to the total number of community codes within the interviews

5.3   PISEC and Physics CoP

In order to determine if participation in PISEC has an impact on UEs physics identity we then look at the intersection of the two communities in the identity codes. That is, when assigning the identity codes to a segment, as part of the operationalization of the framework, we also assign the corresponding community code. A segment could be coded with one identity code and one or more community codes depending on the narrative. There were instances in which even though the experience happened on one CoP, it had an impact on another CoP. In this section, we focus on the intersection of PISEC and physics, more specifically, we identify the cases in which Eoin and Sorcha identities' codes demonstrate an intersection in their membership of PISEC and physics communities. For example, when Eoin is asked about his motivation for joining PISEC he responded *It seemed like a great- I already knew, I'd been working in a lab for a little while already before I joined the first semester, and I already knew how nice it would be to have kind of a break for a few hours every week, and how it would be kind of fun to just interact with the kids and do some cool physics.* This segment was coded as NM-in because he is reflecting on how his membership in the PISEC CoP would positively impact his membership in the physics CoP. Eoin expressed that taking the time off from his physics research in the laboratory to do physics with the children would be beneficial.

Another instance of the intersection between PISEC and physics in Eoin's interview is when he is asked about the impact on the children. In his reflection, he makes a clear affirmation that he is a member of the physics community and that as a member he is actively seeking to engage others in the physics community and he is able to do this, in part, through his participation in PISEC. *We're very happy to get the older brother. 'We', the scientific community, the dark side, are trying recruit them. (laughter) But the younger brother is the kind of person we're trying to get through to it seems like. Clearly he has the aptitude, and maybe if we can show him that then he would be more interested in doing it* (This was coded as NM-inbound).

There are other instances in Eoin's interview where he is reflecting about the connections between informal physics as a broader CoP, PISEC and how those intersects with his physics identity *But yeah, kind of the more I've grown as a scientist, the more I've wanted to help others get into science. I like teaching but I didn't TA at all, so I kind of thought this would be a good alternative.* In his narrative, Eoin often discusses the informal physics community more broadly, rather than the PISEC CoP specifically, with respect to his physics community membership. This result shows in the analysis of his interviews that there is a larger intersection between the physics community and the informal community.

We can also find evidence of the intersection of identity codes between PISEC and physics CoP in Sorcha's interview. For instance, when asked whether she had hesitations to participate in PISEC, Sorcha replied, *My only real hesitations weren't related to PISEC necessarily, they were more related to you know being able to escape the lab. And that's really just a laboratory politics thing, that really doesn't have much to do with PISEC.* This was coded as NM-neutral, because Sorcha is considering how her physics membership would be impacted if she decided to join the PISEC CoP. It was coded neutral because from the narrative we cannot infer if the decision of participation had an effect on her physics identity. She continues the narrative to indicate *You know, the time lost in the lab I could always make up later.* This segment was coded as NM-inbound, because she continues to weigh in on how her participation in PISEC will affect her participation in physics; however, in this segment is clear that she sees participation in PISEC worth it and furthermore that it will not affect her membership in the physics CoP.

Later on, Sorcha discusses different interactions with members of the PISEC CoP that intersect with her physics identity. For example, in two separate occasions she mentions how discussing her physics research with the children in PISEC and having positive feedback from them are experiences that reassure her identity in physics. See quotes below.

> You know, just kind of- One of the girls told me I had the coolest job ever, which again, like when you're drudging through grad school is really fun to hear. (laughter) You're like oh yeah, I totally do! Never mind nothing works. (NE-inbound)

> So it's really cool, and that's always one of my favorite parts is showing the kids around the lab. I just, I love that because they ask all these questions and I just eat it up. So I don't know, it's been really neat for kids to be like 'wow, that's really cool!' It made me feel like I'm actually doing something. (NE-inbound)

> In a sense I think the biggest benefit that I've gotten is, you know, I don't want to say the ego boost, but kind of you know having kids come up to you and be like 'wow, this is really cool!' Because I've given actually- Okay, to back up, I've given lab tours for PISEC in the past, but I never actually did PISEC. (CM-inbound)

Finally, another cases of intersections in Sorcha's interview were related to what she believes is the impact she is having while participating in PISEC. She believes that by doing PISEC and becoming a physics role model for the children she would be able to inspire them to do physics (or science) and show them that anyone can do physics/science, which coincidentally was something she wished she had when she was their age. See quotes below

> But you know, as a whole you could kind of build a rapport and in a sense be like 'I'm a normal human being, kind of nerdy, but normal. And you can too.' (CM-neutral)

> But I'd like to think that there were a couple of them, you know, hopefully a couple of them that not only have like the interest but you know kind of the- Because there is, you know, admittedly a bit of scholastic aptitude that you have to have. Hopefully they get that, you know. (CM-inbound)

The ability to not only convey her physics membership but also share her physics experiences with the PISEC students, allowed Sorcha to increase her level of membership within the PISEC community. The CoP framework allows us to capture these instances of physics and PISEC community intersection as well as many other intersections among identity and community membership.

6 Discussion and Implications

In this paper, we used interviews of two physics PhD students, at different levels in their career, to determine if the CoP framework could capture the nuances in how experiences and interactions, this students had affected their participation within the physics and PISEC CoP. Preliminary analysis indicates that the CoP framework is a plausible tool for studying how participation in informal physics programs impacts university students' development of a physics identity.

Preliminary analysis indicates that the CoP framework is able to capture the differences in Sorcha's and Eoin's participation, and therefore membership, in the PISEC CoP; i.e., both individuals do not participate in the same way in these communities. It also captures their participation in other communities and finally the interaction of their membership of multiple communities. Through the identity codes and subcodes, we are able to identify how experiences in the PISEC community may have had an impact on their physics identity. For Sorcha and Eoin, the social categories of the PISEC CoP were relevant to fostering their identity. Their responses focus on interactions with other members of the PISEC CoP, clearly indicating their experience in that community was influential to their physics identity and that their physics identity was reified through this interaction; i.e., that their membership within PISEC and physics community was positively impacted by interactions with other members of the PISEC CoP.

This is a preliminary analysis, and we are only claiming that the CoP framework is a useful tool for the study of discipline-based identity development through participation in informal programs. Furthermore, identity is a dynamic construct, and therefore, participants may feel very strongly about a community and then change slightly. Different environments and experiences affect our engagement with the community; therefore, changes are expected.

References

Brandt CB (2008) Discursive geographies in science: space, identity, and scientific discourse among indigenous women in higher education. Cult Sci Edu 3(3):703–730

Carlone HB, Johnson A (2007) Understanding the science experiences of successful women of color: science identity as an analytic lens. J Res Sci Teach 44(8):1187–1218

Close EW, Jessica C, Hunter GC (2016) Becoming physics people: development of integrated physics identity through the learning assistant experience. Phys Rev Phys Educ Res 12(1)

Collins (2002) Patricia Hill. Black feminist thought: Knowledge, consciousness, and the politics of empowerment. Routledge

Fracchiolla C, Hyater-Adams S, Finkelstein N, Hinko K (2016) University physics students' motivations and experiences in informal physics programs. In: Paper presented at physics education research conference 2016, Sacramento, CA, 20–21 July 2016

Greenhow C, Robelia B (2009) Informal learning and identity formation in online social networks. Learning Media Technol 34(2):119–140

Hazari Z, Sadler PM, Sonnert G (2013) The science identity of college students: exploring the intersection of gender, race, and ethnicity. J CollSci Teach 42(5):82–91

Holland D, Lachicotte W, Skinner D, Cain Carole (1998) Agency and identity in cultural worlds. Harvard, Cambridge, MA

Hyater-Adams S, Fracchiolla C, Finkelstein N, Hinko K (2018a) Critical look at physics identity: an operationalized framework for examining race and physics identity. Phys Rev Phys Educ Res 14(1):010132

Hyater-Adams S, Finkelstein N, Hinko K (2018b) Performing physics: an analysis of design-based informal steam education programs. In: Paper presented at the physics education research conference 2018, Washington, DC, 1–2 Aug 2018

Irving PW, Sayre EC (2015) Becoming a physicist: the roles of research, mindsets, and milestones in upper-division student perceptions. Phys Rev Spec Topics-Phys Educ Res 11(2):020120

Kearney C (2016) Efforts to increase students' interests in pursuing mathematics, science and technology studies and careers. National measures taken by 30 countries—2015. Report, European Schoolnet, Brussels, Belgium

Lewis KL, Jane GS, Noah DF, Steven JP, Akira M, Geoff LC, Tiffany AI (2017) Fitting into move forward: Belonging, gender, and persistence in the physical sciences, technology, engineering, and mathematics (pSTEM). Psychol Women Quart 41(4):420–436

McGee EO (2016) Devalued black and latino racial identities: a by-product of STEM college culture? Am Educ Res J 53(6):1626–1662

McGee E, Bentley L (2017) The equity ethic: black and latinx college students reengineering their STEM careers toward justice. Am J Educ 124(1):1–36

Miner KN, Walker JM, Bergman ME, Jean VA, Carter-Sowell A, January SC, Kaunas C (2018) From "her" problem to "our" problem: using an individual lens versus a social-structural lens to understand gender inequity in STEM. Ind Organ Psychol 11(2):267–290

National Research Council (2009) Learning science in informal environments: people, places, and pursuits. National Academies Press

Pattison SA, Ivel G, Smirla R-M, Lauren M (2018) Identity negotiation within peer groups during an informal engineering education program: the central role of leadership–oriented youth. Sci Educ

Peña-López I (2016) PISA 2015 results (vol I). Excellence and equity in education

Prefontaine B, Fracchiolla C, Vasquez M, Hinko K (2018) Intense outreach: experiences shifting university students' identities. In: Paper presented at the physics education research conference 2018, Washington, DC, 1–2 Aug 2018

Rahm, J (2007) Urban youths' identity projects and figured worlds: case studies of youths' hybridization in science practices at the margin. Chicago, IL: The Chicago Springer Forum. Science Education in an Age of Globalization

Tellhed U, Bäckström M, Björklund F (2017) Will I fit in and do well? The importance of social belongingness and self-efficacy for explaining gender differences in interest in STEM and HEED majors. Sex roles 77(1–2):86–96

Wenger E (2010) Communities of practice and social learning systems: the career of a concept. In: Social learning systems and communities of practice, pp 179–198. Springer, London

http://www.informalscience.org/evaluation

Enhancing the Teaching and Learning of Physics at Lower Second Level in Ireland

Deirdre O'Neill and Eilish McLoughlin

Abstract The numbers of students studying physics at upper second level and the low numbers of teachers qualified to teach physics at second level are a matter of concern for the future of STEM education in Ireland and internationally. This study describes a national collaboration focussed on addressing teacher's approaches to the teaching and learning of physics at lower second level, raising awareness of physics/STEM careers and examining teachers' and students' academic resilience and unconscious biases. The findings report on the design and implementation of a physics teacher professional development programme with all lower secondary science teachers from seven Irish second-level schools.

1 Introduction

In Ireland, teachers are required to register with the Teaching Council (STEM Education Review 2016), the professional standards body for the teaching profession that promotes and regulates professional standards in teaching. Science teachers register to teach their final degree subject, e.g., Physics, Chemistry, Biology, to upper second level (referred to as Leaving Certificate) and are also recognised to teach science at lower second level (referred to as Junior Cycle science), to students aged 12–15 years. It should be noted that teachers are often qualified to teach one or more subjects depending on their final degree. For example, in 2017, the Teaching Council of Ireland reported that 3878 teachers were registered to teach Biology, 2376 registered to teach Chemistry and 1259 were registered to teach Physics (STEM Education Review 2016). This distribution of science subject specialisms indicates that the majority of Irish students are introduced to physics at Junior Cycle by a non-specialist physics teacher.

D. O'Neill · E. McLoughlin (✉)
CASTeL, School of Physical Sciences, Dublin City University, Dublin 9, Ireland
e-mail: Eilish.McLoughlin@dcu.ie

© The Editor(s) (if applicable) and The Author(s), under exclusive license to Springer Nature Switzerland AG 2020
J. Guisasola and K. Zuza (eds.), *Research and Innovation in Physics Education: Two Sides of the Same Coin*, Challenges in Physics Education,
https://doi.org/10.1007/978-3-030-51182-1_19

239

The 2018 statistics reported by the State Examinations Commissions show that 13% (7535) of the Irish Leaving Certificate student cohort choose to study and complete the physics examination at upper second level. Only 27% (2075) of those studying physics are girls (State Examinations Commission 2018). Worryingly, 22% of the 723 Irish second-level schools do not offer Physics as a separate subject at upper second level (Education 2018). This data highlights the need to concentrate on encouraging more students, particularly girls, to continue in physics at upper second level and encourage more physics students into the teaching profession.

Addressing the low uptake of students studying Leaving Certificate physics is a multifaceted issue that requires a holistic solution. Science education research has put much consideration into developing partnerships between researchers, teacher educators and in-service teachers. Penuel and Gallagher (2017) refers to a mutualistic relationship as a research practice partnership in which the aims of the collaboration are recognised and decided on by both the researcher and educators involved in the partnership. Different types of partnerships for collaboration might include: Networked Improvement Communities, Design-Based Research Practice partnerships, Research Alliances and other Hybrid forms of partnerships (Penuel and Gallagher 2017).

Of central importance for successful collaborations is the development of teacher's pedagogical content knowledge (PCK). Going beyond the knowledge of subject matter and considering its teachability, according to Shulman's model of PCK (Shulman 1986), is paramount to the effective teaching of any subject. Building on this, Etkina (2010) highlights five aspects of PCK that bridges the gap between content and pedagogy in the teaching of physics: (i) orientation towards teaching, (ii) physics curriculum, (iii) student ideas, (iv) effective instructional strategies, (v) assessment methods.

2 Methodology

The new Junior Cycle integrated science curriculum is being rolled out in Ireland since 2015 (National Council for Curriculum and Assessment 2015). The science specifications "allow teachers to employ a variety of teaching strategies depending on the targeted learning outcomes, the needs of their students, and their personal preferences". Student-led inquiry forms the basis of science process as well as developing the content knowledge needed to understand phenomenon (National Council for Curriculum and Assessment 2015).

This study adopts a holistic approach to address shortcomings in physics education and engagement at lower second level in Ireland, and further details of this approach are described in an earlier publication (O'Neill et al. 2018). Seven second-level schools with a total teaching staff of 405 teachers, of which 51 are science teachers, and a total student population of 5149 students (3078 girls, 2071 boys) are partners in this study. These schools are a mix of urban and rural locations based across the greater Dublin area and were selected to be representative of the wider cohort of Irish second-level schools and include two all-girls schools, four co-education schools and two designated disadvantaged schools. The three key objectives of this holistic approach are to:

I. Deepen science teacher's confidence and content knowledge for teaching physics,
II. Increase awareness of STEM and careers in STEM,
III. Adopt a whole school approach to addressing unconscious bias and gender stereotyping and build confidence and resilience for students, particularly girls, to continue with Physics.

This study will report on the approach adopted to address the first and second objectives which focus on deepening science teacher's confidence and content knowledge for teaching physics at Junior Cycle and increasing their awareness of STEM careers. This involved collaboration with fifty-one second-level science teachers of varying tenure and science (Physics, Chemistry, Biology) from across the seven partner schools. Each school has, typically, only one teacher qualified to teach physics at upper second level. The sharing of teaching approaches and classroom practices between science teachers is limited, if it occurs at all.

3 Design Parameters

A series of initial meetings were held with the science teachers to identify their areas of concern when teaching the physics component at Junior Cycle. Teachers discussed areas which their students found difficult in Junior Cycle physics and areas that they (as teachers) would like more support for in teaching Junior Cycle physics. Key topics for workshops were identified, by the science education research team (authors), and followed an open and guided inquiry-based approach (Bevins and Gareth 2018). The workshops focused on specific target concepts where teachers were facilitated to participate as learners and evaluate the learning of these workshops through post-workshop reflections following Guskey's model of Teacher Professional Development (TPD) (Guskey 2002). The target concepts were aligned with the Junior Cycle science curriculum to address one or more of the 5 strands of the Junior Cycle science framework (National Council for Curriculum and Assessment 2015). Table 1 presents an overview of the initial three workshops that were designed and implemented with the science teachers in all of the seven schools which focus on the concepts of light, energy and speed.

The workshop facilitation was school-based and continual across the academic year and scaffolded to develop the five aspects of PCK (Etkina 2010). Teachers participated and reflected on the learnings of each workshop and suggested strategies of implementation into normal classroom practice. During workshops, teachers worked in small groups (2–3) with the longer-term aim of building a sustainable professional learning community (PLC) among the science teachers in each school. Once a year, a university-based workshop was facilitated for all science teachers from the seven

Table 1 Overview of target concepts for science teacher workshops

Workshop	Modules	Target concepts
Light	Light signals and fibre optics Lenses and telescopes	Defining and describing the properties of light Differentiating between absorbing, scattering, reflecting and transparent materials Investigating reflection of light Explaining how optical fibres guide light Describe how lenses focus light Understand the physical concept of "focal point" Identifying real and virtual images Building two types of telescopes Calculating the magnification of a telescope
Energy	Energy and sustainability Heat transfer	Principle of conservation of energy Energy changes Energy efficiency/dissipation Conduction and convection—heat as a form of energy Measure energy inputs/outputs Calculate efficiency Sustainability issues Ethics surrounding the consumption of electricity The effects of global warming
Speed	Introduction to the concept of speed	Planning investigations Developing hypotheses Working collaboratively Identifying variables Forming conclusions
	Relationship between distance, time and speed	Forming coherent arguments Graphical representation of data Interpretation of scientific data

schools to come together to share practices and focus on improving the student's ability to learn (Dana and Yendol-Hoppey 2015).

4 Workshop Summary

The school-based workshops were typically 90 min long and delivered to all science teachers in each school with between four and eleven teachers at each workshop. Firstly, the context of the workshop was introduced by the facilitator (usually lead author) with the whole group using a variety of approaches, e.g., open questions, careers and applications, preconceptions and understanding, of a topic, pedagogical approach, historical context and phenomenon. A series of investigations were set up around the laboratory to address target concepts of the workshop. Teachers were facilitated in small groups (2–3) to carry out investigations with associated worksheets. Whole group discussions were facilitated at various intervals (typically 3–4 times during workshop) to facilitate reflected on: teaching and learning strategies, areas of improvement and assessment and opportunities for classroom implementation.

5 Light Workshop

The focus of the light workshop was to facilitate teachers in developing their understanding of light through two modules: (i) Lenses and telescopes (ii) Light signals and fibre optics. The EYEST photonics kit developed by VUB, B-PHOT Brussels Photonics (Prasad et al. 2012) was used providing examples of investigations that could be used in classroom practices and complement existing secondary school science curriculum. The context of the workshop was set by highlighting with teachers the difficulties students have interpreting some of the concepts within the topic of light. Teachers reviewed a list of these difficulties and preconceived ideas as shown in Table 2 (Sampson and Schleigh 2013; Kaewkhong et al. 2010), and they identified the most common statements they heard from their students.

The first module began with a group discussion to identify different modes of communication. Three activities with a worksheet were used to develop key learnings with a focus on light as a method of communication before students (teachers acting as students) were required to finally plan and carry out an open inquiry. The second module used eight guided inquiry activities centred around the concept of the refraction of light. At the end of the worksheet, a whole group discussion was facilitated to discuss and falsify the statements of conceptual difficulty for students, using scientific argumentation based on the target concepts investigated in both modules.

Relevant careers associated with the topic of light were used as a resource within the workshop with emphasis on job description and transferrable skills required (GradIreland 2018). Workshop activities also included strategies to enable teachers to discuss methods of encouraging students to research a career they had not heard of before on the list in order to increase their awareness of careers in the field.

6 Energy Workshop

The energy workshop was adapted from the *Energy in Action* activities for Junior Cycle published by the Sustainable Energy Authority in Ireland (SEAI) (Energy in Action Sustainable Energy Authority of Ireland 2019). This topic enabled teachers to draw on students' prior knowledge about energy. The approach taken for this workshop encompassed the cross-cutting themes of energy as described in the Junior Cycle science specifications and aimed to foster a deeper understanding of energy

Table 2 Statement of student conceptual difficulty in light (Schleigh 2014)

• Different wavelengths of light have different energy and therefore different speeds	• The size of the image depends on the diameter of the lens
• The distance light travels depend on day or night	• Light needs air to travel
• Black does not reflect any light and/or white does not absorb any light	• The distance that light travels depends on the amount of energy that light has
• Only shiny materials reflect light	• Objects that reflect are sources of light (e.g., the Moon)
• Water does not reflect or absorb light, but light can go through it	• When a lens has moved an image will become bigger or smaller but will always remain sharp
• Sunlight is hot (has energy) but visual light is not	• Our eyes produce light so we can see things
• The size of the image depends on the diameter of the lens	• A radio wavelength is a sound wave not part of the electromagnetic spectrum
• The stronger the source of light the bigger the shadow and the bigger the source of light the smaller the shadow	• Moving position when looking at a mirror image will change the amount of the image that can be seen or the position of the image in the mirror
• Shiny objects reflect more light than dull objects	• A shadow is a reflection from the Sun
• Light always passes straight through transparent objects (without changing direction)	• An observer can see more of himself by backing up
• Shadows are always black	

in everyday context and the abstract and mathematical idea of energy as proposed by Millar (2005). Teachers considered the scientific meaning of energy, energy in everyday contexts, arguments about teaching energy and teaching energy ideas.

The target concepts of this workshop were addressed through teachers carrying out guided inquiry experimental investigations. Teachers working in small groups moved around to different stations investigating the target concepts and the facilitator raised questions relating to student learning and understanding of each concept. Classroom dialogue was focussed on the assessment strategies used in this workshop. Teachers reflected on the workshop highlighting areas of improvement and content knowledge difficulty that needs to be considered in classroom implementation.

STEM career awareness in the subject area of energy was listed as a resource identifying transferrable skills, top employers and job roles (Grad Ireland 2018). Similar to the light workshop, a short whole-group discussion was facilitated to suggest appropriate career investigation promoting STEM awareness.

7 Speed Workshop

The topic of speed was approached using the speed unit from the *SAILS Inquiry and Assessment Units* (Harrison 2014). This teacher education unit was designed to develop scientific skills, in particular; hypothesising, forming coherent arguments and working collaboratively. In addition, students developed their scientific reasoning and scientific literacy. The context of this workshop focussed on classroom assessment strategies based on teacher observations, classroom dialogue, evaluation of student artefacts and self-assessment. Teachers role-played a classroom scenario of groups of students (2–3) using both an open and guided inquiry approach with one of the participants enacting the role of the teacher to focus on assessment methods. Whole group discussions took place after teachers had completed the open-guided inquiry activities. This allowed teachers to discuss the advantages and disadvantages of both inquiry approaches from both the student and teacher perspective.

As a final activity, teachers built on their scientific argumentation skills developed in this workshop and the previous workshop on energy to analyse and explain graphing skills and representations in physics. Teachers completed a reflection tool at the end of the workshop which formed part of the evaluation of the study.

8 Discussion

The design and implementation of the series of teacher education workshops used in this study and focussed on different aspects of pedagogical and organizational knowledge for teaching physics at Junior Cycle. The three workshops described collectively address the five aspects of PCK for teaching physics as identified by Etkina (2010):

(i)   orientation towards teaching: light, energy
(ii) physics curriculum: light, energy, speed
(iii) student ideas: light, speed
(iv) effective instructional strategies: energy, speed
(v)   assessment methods: speed.

Overall, the design of the workshops focused on deepening science teacher's confidence and content knowledge for teaching physics at Junior Cycle and increasing their awareness of STEM careers (objectives I and II). It has been observed that within each workshop very specific focus on the subject matter was needed to deepen teacher's physics pedagogical content knowledge on that topic. This finding has been reported by Etkina et al. (2018) and references therein, that the teachers of a specific subject (e.g., physics) need to possess knowledge that is different from the knowledge of other content experts (e.g., a physicist). The concept of CKT originated with the pedagogical content knowledge (PCK) work of Shulman.

CKT is premised on the idea that teachers need to understand subject matter knowledge in ways that are specific to teaching, such as understanding the historical foundations of the concepts that students need to learn, structure of the curriculum that allows students to build coherent understanding, challenges that specific subject matter knowledge might present to students and how students may represent their understanding in nonstandard forms, knowing what knowledge representations are helpful, how to ask questions or provide explanations that can move understanding forward, etc. (Etkina et al. 2018)

These findings highlight the importance of the integration of the CKT model in the design and implementation of physics teacher professional development programmes. The next phase of this study will focus on the evaluation of this holistic approach on teacher's CKT in physics as well as student learning and understanding in physics. Future workshops with science teachers will also incorporate the wider objective of adopting a whole school approach to addressing unconscious bias and gender stereotyping and aims to build confidence and resilience among students, particularly girls, to continue with Physics (objective III). This is an important aspect of this study as Hazari et al. (2010) emphasises that students' physics identities are shaped by their experiences in second-level physics classes and by their career outcome expectations.

References

Bevins S, Gareth P (2018) Reconceptualising inquiry in science education. Int J Sci Educ 38:17–29. https://doi.org/10.1080/09500693.2015.1124300

Dana NF, Yendol-Hoppey D (2015) The PLC book. Corwin Press

Education ie. (2018) Available from: https://www.education.ie/en/Publications/Education-Reports/STEM-Education-in-the-Irish-School-System.pdf. Accessed 8 May 2018

Energy in Action Sustainable Energy Authority of Ireland (2019). Available from: https://www.seai.ie/teaching-sustainability/post-primary-school/energy-in-action/. Accessed 13 Jan 2019

Etkina E (2010) Pedagogical content knowledge and preparation of high school physics teachers. Phys Rev Spec Top Phys Educ Res. 6:020110. https://doi.org/10.1103/PhysRevSTPER.6.020110

Etkina E, Gitomer D, Iaconangelo C, Phelps G, Seeley L, Vokos S (2018) Design of an assessment to probe teachers' content knowledge for teaching: an example from energy in high school physics. Phys Rev Spec Top Phys Educ Res 14:010127. https://doi.org/10.1103/PhysRevPhysEducRes.14.010127

GradIreland (2018) Available from: https://gradireland.com/. Accessed 13 Jan 2019

Guskey TR (2002) Professional development and teacher change. Teachers Teach 8:381–391. https://doi.org/10.1080/135406002100000512

Harrison C (2014) Assessment of inquiry skills in the sails project. Sci Edu Int 25:112–122

Hazari Z, Sonnert G, Sadler PM, Shanahan MC (2010) Connecting high school physics experiences, outcome expectations, physics identity and physics career choice: a gender study. J Res Sci Teach 47:978–1003. https://doi.org/10.1002/tea.20363

Kaewkhong K, Mazzolini A, Emarat H, Arayathanitkul K (2010) Thai high-school students' misconceptions about and models of light refraction through a planar surface. Phys Educ 45:97. https://doi.org/10.1088/0031-9120/45/1/012

Millar HR (2005) Teaching about Energy. (ISBN: 1 85342 626 1)

National Council for Curriculum and Assessment (2015) Junior cycle science curriculum specification. Available from: https://www.curriculumonline.ie/Junior-cycle/Junior-Cycle-Subjects/Science. Accessed 13th Jan 2019

O'Neill D, McLoughlin E, Gilheany S (2018) Creating research-practice collaborations to address gender imbalance in physics at second level. In: Proceedings of the eighth science and mathematics education conference connecting research, policy and practice in STEM education pp 67–72

Penuel WR, Gallagher DJ (2017) Creating research practice partnerships in education. ERIC

Prasad A, Debaes N, Cords N, Fischer R, Vlekken J, Euler M, Thienpont H (2012) Photonics explorer: revolutionizing photonics in the classroom. Int Soc Optics Photonics. https://doi.org/10.1117/12.979339

Sampson V, Schleigh S (2013) Scientific argumentation in biology: 30 classroom activities. NSTA Press. Available at: http://static.nsta.org/connections/elementaryschool/201404Schleigh.pdf [Accessed 13th January 2019]

Schleigh S (2014) Assessments in the arguments. Sci Child 51:46

Shulman LS (1986) Those who understand: knowledge growth in teaching. Educ Res 15, 4–14. https://doi.org/10.2307/1175860

State Examinations Commission (2018) State Examination Statistics, Available at: https://www.examinations.ie/statistics/?l=en&mc=st&sc=r12. Accessed 8 May 2018

STEM Education Review Group (2016) A Report on Science, Technology, Engineering and Mathematics (STEM) Education Analysis and Recommendations Available from: https://www.education.ie/en/Publications/Education-Reports/STEM-Education-in-the-Irish-School-System.pdf. Accessed 8 May 2018

Printed in the United States
by Baker & Taylor Publisher Services